# Microelectrode Methods for Intracellular Recording and Ionophoresis

# Biological Techniques Series

J. E. TREHERNE
*Department of Zoology*
*University of Cambridge*
*England*

P. H. RUBERY
*Department of Biochemistry*
*University of Cambridge*
*England*

Ion-sensitive Intracellular Microelectrodes, *R. C. Thomas* 1978
Time-lapse Cinemicroscopy, *P. N. Riddle* 1979
Immunochemical Methods in the Biological Sciences: Enzymes and
    Proteins, *R. J. Mayer* and *J. H. Walter* 1980
Microclimate Measurement for Ecologists, *D. M. Unwin* 1980
Whole-body Autoradiography, *C. G. Curtis*, *S. A. M. Cross*,
    *R. J. McCulloch* and *G. M. Powell* 1981
Microelectrode Methods for Intracellular Recording and Ionophoresis,
    *R. D. Purves*

# Microelectrode Methods for Intracellular Recording and Ionophoresis

**R. D. Purves**

Department of Anatomy and Embryology,
University College London,
London

1981

ACADEMIC PRESS

*A Subsidiary of Harcourt Brace Jovanovich, Publishers*

London New York Toronto Sydney San Francisco

ACADEMIC PRESS INC. (LONDON) LTD
24–28 Oval Road,
London NW1

*U.S. Edition published by*
ACADEMIC PRESS INC.
111 Fifth Avenue
New York, New York 10003

*British Library Cataloguing in Publication Data*

Purves, R. D.
  Microelectrode methods for intracellular
  recording and ionophoresis. – (Biological
  techniques series)
  1. Cytology – Technique     2. Electrodes
  I. Title     II. Series
  QH585.5.E/       574.19'285       80-42080

  ISBN 0-12-567950-5

Typeset by Preface Ltd, Salisbury, Wilts.
**Printed in Great Britain by**
**John Wright & Sons, Ltd., at the Stonebridge Press, Bristol**

# *Preface*

"The resting potential was $-67.21$ mV", "wash the tubing in alcohol", "surface tension prevents filling", "suck the steam off slowly", "the fibre-glass method is too expensive", "disconnect the earth lead at the plug", "use a notch filter", "balance the bridge before impalement", "release is governed by Faraday's Law", "lengthen the probe arm to cushion vibrations" . . . .

Each of the above examples points to a fallacy in the folklore of intracellular recording with microelectrodes. One of them could produce a lethal hazard, others waste time or money, and some merely imply loose thinking. In writing this book I have tried to encourage safe, efficient, soundly-based practices. My approach is to show which methods are good and which are bad, by outlining the principles (physical, electrochemical or electronic) that underlie them. Advice for the novice, as well as the more experienced user of microelectrodes, has been included wherever possible. For example, the practical problems of coping with electrical and mechanical interference are dealt with in considerable detail. Many electronic circuit diagrams are given, and there is an elementary introduction to the relevant parts of electronic theory. I hope that the book will prove equally useful on the laboratory bench and as a reference text.

At various stages during its preparation the manuscript was greatly improved by comments from Ian Hill-Smith, Don Jenkinson, Barbara Pittam, Lutz Pott, Roger Thomas and Luca Turin. I am grateful for their suggestions.

*January 1981*                                                              R. D. P.

# *Contents*

Preface .......................................................... v

**1**
**Introduction to Microelectrodes**
I.    The ubiquitous microelectrode .......................... 2
II.   An overview ............................................ 3
      A.  The microelectrode ................................ 3
      B.  The indifferent electrode ......................... 4
      C.  The preamplifier .................................. 5
      D.  Micromanipulators ................................ 5
      E.  Output devices .................................... 6
      F.  The preparation ................................... 7
      G.  How to impale cells .............................. 8
      H.  What is the resting potential? ................... 9
III.  Other recording methods .............................. 10
      A.  Extracellular recording .......................... 10
      B.  Sucrose gap ...................................... 10
      C.  Kostyuk's method ................................. 11
      D.  Voltage-sensitive dyes ........................... 12

**2**
**Making and Filling Microelectrodes**

I.    Introduction .......................................... 13
II.   Glass ................................................. 13
III.  Pipette pullers ....................................... 14
IV.   Filling solutions ..................................... 15
V.    Filling methods ....................................... 17
      A.  Direct through tip ............................... 17
      B.  Direct through stem .............................. 19
      C.  Boiling .......................................... 19
      D.  Vacuum ........................................... 20
      E.  Alcohol .......................................... 20
      F.  Pre-filled ....................................... 21
      G.  Centrifugation ................................... 21

H.   Pressure ................................................... 21
I.   Distillation .............................................. 22
J.   Local heating ............................................ 22
K.   Vibration ................................................ 23
L.   Mobile filament .......................................... 23
M.   Glass fibre .............................................. 23
VI.  Bending, breaking and bevelling .......................... 24

**3**
**The Physics of Microelectrodes**

I.    Introduction ............................................ 26
II.   Size and shape .......................................... 26
III.  Mechanics ............................................... 28
IV.   Electrical resistance ................................... 28
V.    Capacitance ............................................. 31
VI.   Tip potentials .......................................... 33
VII.  Non-linear electrical behaviour ......................... 35

**4**
**Circuitry for Recording**

I.    Introduction ............................................ 39
II.   Preamplifiers ........................................... 39
      A.  Understanding the specifications .................... 39
      B.  What's inside ....................................... 43
      C.  The driven shield ................................... 45
      D.  All about negative capacitance ..................... 47
III.  Joining wires to cells .................................. 49
      A.  Connecting to the microelectrode ................... 49
      B.  Connecting to the extracellular medium ............. 51
IV.   Offset voltage control .................................. 55
V.    Earthing and interference ............................... 55
      A.  Alternating electric fields ........................ 57
      B.  Alternating magnetic fields ........................ 59
      C.  Earth loops ........................................ 59
      D.  What to do if it hums .............................. 61
      E.  Three case histories ............................... 61
      F.  Stimulus artefacts and impulsive interference ...... 66
VI.   Noise ................................................... 67
VII.  Measuring microelectrode resistance ..................... 71
VIII. Low-pass filters ........................................ 72
IX.   Calibration ............................................. 73
X.    Trouble-shooting and fault-finding ...................... 74

**5**
**Current Injection**

| | | |
|---|---|---|
| I. | Introduction | 76 |
| II. | Current sources | 76 |
| III. | Current monitors | 79 |
| IV. | Alternate current injection and potential measurement | 81 |
| V. | Simultaneous potential measurement | 82 |
| | A. Current pump with electronic subtraction | 83 |
| | B. Current pump with direct cancellation | 83 |
| | C. Old-fashioned bridges | 84 |
| | D. Comparison between methods | 85 |
| | E. How to balance the bridge | 86 |
| | F. Errors and artefacts | 88 |
| | G. Multiplex method | 91 |

**6**
**Ionophoresis**

| | | |
|---|---|---|
| I. | Introduction | 92 |
| II. | The physics of ionophoresis | 92 |
| | A. The transport processes | 92 |
| | B. The relation between current and drug release | 95 |
| III. | The practice of ionophoresis | 98 |
| | A. Circuitry | 98 |
| | B. Intracellular application | 100 |
| | C. Miscellaneous hints | 101 |
| IV. | Other methods of drug application | 102 |

**7**
**Control of Vibration**

| | | |
|---|---|---|
| I. | Introduction | 103 |
| II. | Principles and practice of anti-vibration | 103 |

**Appendix I**
**Electronics for Microelectrode Users**

| | | |
|---|---|---|
| I. | Ohm's law and other circuit theorems | 107 |
| II. | Capacitance | 110 |
| III. | Operational amplifiers | 113 |

**Appendix II**
**Current–Voltage Relation of Microelectrodes**          116

**Appendix III**
**A Simple Preamplifier**          120

**Appendix IV**
**Preamplifier with Current Injection**                        125

**Appendix V**
**A Simple Current Pump**                                      132

**Appendix VI**
**Drug Concentration and Ionophoretic Release**                135

**References**                                                 137

**Index**                                                      145

# 1
# *Introduction to Microelectrodes*

The best way to gain an introduction to microelectrodes is to spend some time in a laboratory where microelectrodes are used. A poor alternative is to invite an experienced person to help you get your own microelectrode laboratory started, or to turn to a work such as the present one. With this in mind, some parts of this book have been written at an elementary level to provide further help for the beginner. The whole of Chapter 1 and most of Chapters 2 and 4 should be immediately comprehensible, although it is hoped that experienced practitioners will not find them offensively oversimplified. Chapters 3, 5 and 6 are of a more specialized nature and could safely be omitted on a first reading. Chapter 7 represents an attempt to inject some theory into what is usually regarded as a commonsense matter; whether or not it will prove useful will depend on the tastes of individual readers.

A number of other books and descriptive articles deal with intracellular microelectrode techniques: Donaldson (1958); Fatt (1961); Nastuk (1964); Bureš, Petráň and Zachar (1967); Geddes (1972); Dichter (1973); Kuriyama and Ito (1975); Thomas (1978). The content of some of these is outdated and none offers a complete account of the theory and practice of modern microelectrode usage. Nevertheless, all give valuable information and hints. The present book duplicates little of the earlier sources and so the reader is advised to consult as many of them as possible.

An important prerequisite for the intending user of microelectrodes is some knowledge of electricity. He or she must understand Ohm's law, be familiar with alternating and direct current, know what a voltage amplifier does, and have at least a vague idea what capacitance is. Deficiencies will be speedily revealed when the exercises of Appendix I are attempted. The conscientious working through of these or similar

1

exercises ought to be part of every electrophysiological training. There are not many electronics textbooks that provide the correct balance for the beginning microelectrode user; most are obsessed with the microphysics of the *pn* junction and the transistor, topics of little or no utility in the present context. The fundamental active device in modern analogue electronics is the integrated operational amplifier. The advent of cheap operational amplifiers of high performance has made electronic design so easy that even non-specialists can plan and construct a wide range of analogue circuitry. Some recommended texts are those by Smith (1971) and Young (1973). Despite its beguiling title, "Electronics for Neurobiologists" by Brown, Maxfield and Moraff (1973) is unsuitable as an introductory text, although it may be found useful for occasional reference.

The beginning user of microelectrodes is often as ignorant of mathematics as of electronics. Although mathematics may appear to have little connection with practical electrophysiology, two books are so outstandingly good that I cannot forgo the opportunity of recommending them. Riggs (1963) offers a "critical primer" of mathematics in physiology, and Stark (1970) gives a superb introduction to numerical methods. Both authors are concerned to teach, and not merely to expound. The textbook of electrochemistry by Bockris and Reddy (1970) is recommended for the same reason.

## I. The Ubiquitous Microelectrode

Intracellular electrical recording with microelectrodes is usually considered to have begun with the measurements of resting potential in frog muscle by Ling and Gerard (1949), but numerous attempts had been made, more or less ineffectually, before that. For example, Hogg, Goss and Cole (1934) recorded positive-going action potentials of about 5 mV amplitude from heart muscle grown in tissue culture. The analytic capabilities of intracellular microelectrode recording were powerfully demonstrated by Nastuk and Hodgkin (1950) in a study of the action potential, and by Fatt and Katz (1951) in a study of transmission at the nerve–muscle synapse. Transmembrane microelectrode recording was rapidly established as the most important single electrophysiological method, and remains so today. The principles of intracellular recording also remain, virtually unchanged, although modern developments have made the method very much easier and more convenient.

As well as measuring membrane potential, Fatt and Katz (1951) used microelectrodes to inject electric current into impaled muscle cells. This second application of microelectrodes now finds employment for stimulating nerve and muscle cells, determining passive electrical properties of cells and their membranes, measuring null ("reversal") potentials of neurotransmitters, and controlling membrane potential under voltage clamp.

Another important use for the microelectrode was devised by Nastuk (1953*b*), who described membrane potential changes of muscle cells "produced by transitory application of acetylcholine with an electrically controlled microjet". The microjet came from the end of a microelectrode containing a solution of acetylcholine. This method, variously called "ionophoresis", "iontophoresis", or "micro-electrophoresis", is one of the fastest ways of applying drugs and transmitter chemicals to cells.

The above three uses of glass microelectrodes form the subject of this book. In passing we may note other ways of exploiting microelectrodes: extracellular electrical recording, micropuncture in renal physiology, pressure recording in small vessels, and (with some modification) measurement of pH and ion concentrations inside and outside cells (Thomas, 1978).

## II. An Overview

A set of apparatus for intracellular recording is schematized in Fig. 1. Some of the items shown will receive detailed treatment later and are merely introduced here.

### A. *The Microelectrode*

Following the usage of Thomas (1978), the fine-tipped tapering glass capillary tube will be called a micropipette (Fig. 2). When filled with electrolyte solution and used for measuring potentials it becomes a microelectrode. Strictly it is not an electrode at all. The real electrode (Chapter 4, Section III) is the junction between metal, usually silver, and electrolyte, usually KCl, in the stem of the micropipette or just outside. The microelectrode is only a micro salt-bridge. To be even more precise, it is what electrochemists call a "half-bridge". The most important single attribute of a microelectrode is its external diameter near the tip. As a rule of thumb this needs to be 0.5% or less of the diameter of the cell which it is to impale. Thus a cell 100 $\mu$m in diameter may be impaled without gross damage by a pipette of 0.5 $\mu$m tip diameter, just large enough to resolve in the light microscope. But

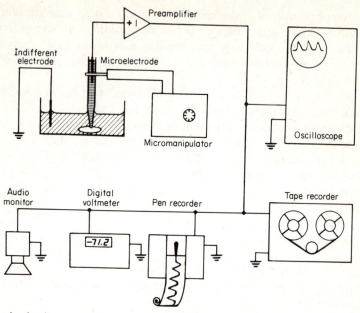

Fig. 1. A simple arrangement for intracellular electrical recording, together with a selection of output devices.

most cells are smaller, and the pipettes used on them have tips which cannot be seen in the light microscope. The electron microscope, especially the scanning variety, can give good pictures of the tip region (Chapter 3, Section II).

## B. The Indifferent Electrode

A voltmeter measures the potential difference between two points, and therefore needs to have two electrical leads. For intracellular recording, the microelectrode forms one lead and the indifferent electrode the other. Commonly the indifferent electrode will be at or near zero potential, as in Fig. 1 where it is shown connected to earth. An alternative name is "bath electrode", which is appropriate for studies on isolated tissues, but less so for studies on whole animals in which the

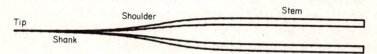

Fig. 2. The parts of a micropipette.

indifferent electrode will probably be a strip of silver buried in the musculature. Indifferent electrodes are discussed in Chapter 4, Section III.

### C. The Preamplifier

This piece of equipment has several important jobs, but despite its name voltage amplification is not necessarily one of them. An alternative name is "head-stage". "Electrometer" amplifier is sometimes used, the epithet having no precise technical meaning. "Cathode follower" was once a fine descriptive term for the preamplifier, but only the most reactionary laboratories continue to use vacuum tubes as input devices. I recall, during my introduction to electrophysiology, having to warm up a pair of ME 1400s an hour before the experiment was due to start. I also remember that they were relegated to a top shelf as soon as the first commercial solid state preamplifier arrived. In fairness, it must be admitted that a good cathode follower can hold its own when electrical robustness and low noise are the chief requirements.

What is the preamplifier for? Obviously it must pass the voltage signal from the microelectrode to the oscilloscope. It may be designed to do this without altering the amplitude of the signal ("unity gain"), or there may be some amplification built in (usually ×5, ×10 or ×100). The gain of a cathode follower is appreciably less than unity, say 0.95, and varies somewhat. The gain of modern solid state circuits should be stable within 1% over many weeks or even years.

The second job of the preamplifier is to prevent any significant current from flowing through the microelectrode. Small neurones may be excited by a current of only 0.1 nA ($10^{-10}$ A), and so the *leakage current* generated by the preamplifier must be much less than that. In addition, the preamplifier must not draw current from the microelectrode when it is recording a cellular potential. If it did, there would be a voltage drop across the resistance of the microelectrode (Ohm's law), and the potential seen by the preamplifier would not be a faithful replica of that seen by the tip of the microelectrode. The performance of a preamplifier in this respect is indicated by its *input resistance*. Preamplifiers are considered in detail in Chapter 4.

### D. Micromanipulators

Little useful advice can be offered about the choice of a micromanipulator from the wide range available commercially. Skilful people can do delicate work with a crude manipulator. Clumsy people, or those working on small cells, need better equipment, and should perhaps contemplate using a manipulator that is controlled remotely by electrical, pneumatic or hydraulic means. A directly controlled micromanipulator needs to be strongly made and securely bolted down if it is not to bend and wobble.

The really important thing is to ensure that your micromanipulator has a fine control of movement *along the axis* of the microelectrode. If the electrode has to be mounted obliquely, this consideration may rule out those manipulators that can only advance the microelectrode vertically.

Micromanipulator design principles are discussed by El-Badry (1963) and Kopac (1964).

### E. Output Devices

Figure 1 shows five typical output devices, not all of which are likely to be in use at the same time. The most versatile is the oscilloscope. If you have sufficient funds to buy only one output device, an oscilloscope should be your first choice unless you know you will not want to record rapid events.

*1. The oscilloscope.* In all respects but one, the performance of oscilloscopes is better than is needed for electrophysiological purposes. There are few pitfalls in the choice of an oscilloscope; almost any model will serve. In the simpler kinds the displayed traces are evanescent; if not photographed at the time or retained by an instrumentation tape recorder they are gone forever. The bistable storage screen introduced by Tektronix Inc. some years ago captures the traces which can then be studied and photographed at leisure. Few workers who have used a storage oscilloscope would willingly do without one. A more recent innovation offering greater flexibility is digital storage.

Cheap oscilloscopes, or cheap plug-in modules for expensive oscilloscopes may have insufficient amplification to display small signals such as "miniature" synaptic potentials. This problem could of course be overcome by acquisition of a more expensive oscilloscope or a preamplifier with built-in voltage amplification. It could also be solved with trivial expense by interposing an operational amplifier circuit of fixed gain ×10 between the preamplifier and the oscilloscope. An electronics workshop will produce this within a few hours or days if given a clear statement of what is wanted. Alternatively, the construction of the circuit might form a useful introduction to electronic practice for the experimenter himself (see Exercise 20 of Appendix I).

*2. The tape recorder.* Despite the inroads made by computer-based data storage techniques, an FM tape recorder remains a highly useful piece of equipment for retaining experimental results that are to be analysed later. "FM" stands for "frequency modulation" and is a way of making magnetic tape remember slowly changing signals at the expense of its ability to record rapidly changing ones. An ordinary domestic tape recorder is only suitable for signals which contain no relevant information at low (<40 Hz) frequencies. Unfortunately, intracellularly recorded signals usually do contain low frequency information — the resting potential is an example — and so a relatively expensive instrumentation FM tape deck is required.

A voice channel is a desirable feature which allows a commentary to be recorded alongside the data and relieves the experimenter of the need to make copious written notes.

*3. The pen recorder.* Unlike oscilloscopes, pen recorders provide instant "hard copy". Distinguish carefully between an ordinary "galvanometric" pen recorder and a servo-controlled "potentiometric" chart recorder. The former writes on narrow paper and its measurement accuracy is often worse than 5%. The latter writes on wide paper with an accuracy of about 2% or better, but has a much slower response.

*4. The digital voltmeter.* Formerly a luxury because of its high price, the digital voltmeter (DVM) or its less versatile cousin the digital panelmeter (DPM) is now so cheap that every laboratory should have one. The so-called $3\frac{1}{2}$ digit types, with a maximum numerical reading of 1999, provide enough precision for physiological work. The main use of a DVM is in the continuous monitoring of resting potentials (Chapter 4, Section IV).

*5. The audio monitor.* The audio monitor provides you with an audible indication of the resting potential and lets your ears do some of the work of deciding when a cell has been impaled. Essentially it consists of a voltage-controlled oscillator and loud-speaker. It is better if the circuit is made in such a way that the pitch of the emitted note rises on impalement. Cell death and decline of the resting potential are then signalled by a mournful fall in pitch.

### F. The Preparation

Since the aim of this book is to teach you about the hardware of intracellular recording, especially those items in Sections A, B and C above, the tissue preparation will make few appearances. But a good grasp of the principles of microelectrode usage should free you to think more about the preparation and its physiology, and less about the technical aspects.

Try not to think of the preparation as a featureless grey blur, as it may very likely seem when viewed through a dissecting microscope. Look up its histology in a textbook, and study electronmicrographs. Investigate the feasibility of using a compound microscope, if the preparation is thin enough. Try brightfield, darkfield, phase contrast and differential interference contrast (Nomaski) optics. Is it possible to do your experiment with proper microscopic viewing? If so, should you use an inverted microscope (objective lenses don't get in the way) or an upright one (better visualization of the top layer of cells)? Either way, use a microscope with tube focusing; a stage focusing microscope may break the tips of your microelectrodes when you adjust it.

## G. How to Impale Cells

Practise on easy tissues, with the absolute minimum of apparatus. The frog's sartorius muscle or the mouse's diaphragm are good starters. When you get bored with measuring resting potentials put a pair of wires on the muscle and stimulate it. Try to record action potentials. One or two days' work on these easy tissues should encourage you to attempt something more adventurous, such as the isolated atria of the mouse heart.

If the cells are big and the microelectrode sharp, and there are no tough connective tissue barriers, just advance the microelectrode until a resting potential is registered. Wait a few seconds to see whether the resting potential improves. If it doesn't, try cautious withdrawal or advancement of the microelectrode; this sometimes improves the quality of the impalement.

Experiment with the various ways of "popping" an electrode in. Tap gently on the baseplate with your fingers, using a staccato action, as you advance the electrode with the other hand. If the connective tissue is very tough, a tap on the micromanipulator itself may do the trick. Try tapping the baseplate with a screwdriver to get a staccatissimo action. Mechanical devices for advancing the microelectrode rapidly through the cell membrane are discussed by Brown and Flaming (1977). If your preamplifier has negative capacitance, turn the capacitance control to maximum for a fraction of a second, to make the amplifier oscillate. When the electrode tip is dimpling the surface of a cell, the oscillation acts as a micro-cautery and often allows impalement. A push-button control can be fitted to such an amplifier to simplify the manoeuvre (Appendices III and IV). Inevitably, but regrettably, the control is known as a "zap" button.

If you have to maintain impalements for long times, pay great attention to the mechanical design of your set-up (Chapter 7). Make every attempt to immobilize the preparation. Isolated tissues need to be well anchored in the organ bath with the aid, for example, of lots of entomological pins. A 2–4 mm layer of Dow Corning "Sylgard" silicone plastic makes a good floor for the organ bath, being transparent and able to hold pins well.

A common problem in recording from small cells is a rapid decline in resting potential following an initially satisfactory impalement. When all else has failed try increasing the extracellular calcium concentration to two or three times its usual value. This sometimes has a dramatic effect on the quality of impalements. It may, however, make it difficult to get the results published!

### H.  What Is the Resting Potential?

Resist the temptation to report resting potentials with unrealistic precision, e.g. $-71.28$ mV, even if you have a digital voltmeter which shows that reading momentarily after impalement. The above value should be reported as $-71$ mV or, more usually "about $-70$ mV". Read the paper by Tasaki and Singer (1968) every year or two to immunize yourself against the idea that resting potentials can be measured accurately.

The apparent resting potential is the difference between the potentials measured when the microelectrode is inside and outside the cell. For a variety of reasons this may not be very close to the "true" resting potential. Damage inflicted by the microelectrode is an obvious source of error. There are in addition two systematic errors which arise from the alteration in chemical environment of the electrode tip as it enters the cell. The first is a change of the *liquid junction potential* (Chapter 2, Section IV) between the microelectrode's filling solution and the electrolyte outside the tip. The second is a change of the *tip potential* (Chapter 3, Section VI).

The magnitude of these systematic errors is difficult to estimate. The usual practice has been to ignore the change of liquid junction potential on the grounds that it ought to be very small if the microelectrode contains the usual 3 M KCl filling solution, although a recent study by Hironaka and Morimoto (1979) suggests that the error may amount to several millivolts. We saw earlier that a microelectrode is an example of a salt bridge. According to Ives and Janz (1961) measurements made with salt bridges are liable to "an absolute uncertainty of 1 to 2 mV".

These strictures do not necessarily apply to changes of membrane potential; there are many instances where a change of 100 $\mu$V is both measurable and relevant.

### III. Other Recording Methods

Remember that intracellular microelectrodes do not have a monopoly of the recording of electrical signals from cells. Some alternatives are mentioned briefly here.

### A. Extracellular Recording

Easy to do but difficult to interpret, extracellular recording is a method in some ways complementary to intracellular recording. Figure 3(a) and (b) shows that the two methods need not be different in principle. The figure also explains why a monophasic action potential recorded extracellularly is negative-going, a fact which puzzles many students. Extracellular recording is described in many texts. See for example Delgado (1964), Donaldson (1958), and Strong (1970).

### B. Sucrose Gap

The amplitude of signals recorded extracellularly is usually only a small fraction of the true transmembrane amplitude, because the signal is shunted by low-resistance extracellular pathways for current flow. If the pathways are obstructed (as shown in Fig. 3(c)), the recorded signal gets bigger. Any substance of high specific resistivity can be used: air, paraffin oil, and the glass tube of a suction electrode are among the commonest. A solution of very pure sucrose in deionized water has the advantage that it can penetrate between cells and thereby increase extracellular resistance to extremely high levels. The sucrose "gap" is thus a gap in the extracellular pathway for current flow (Fig. 3(d)). Enthusiasts for the sucrose gap claim that in suitable preparations it combines the advantages of intra- and extracellular recording. Technical descriptions may be found in references given by Boev and Golenhofen (1974) and Coburn, Ohba and Tomita (1975), but the method is best learned from someone who has experience with it.

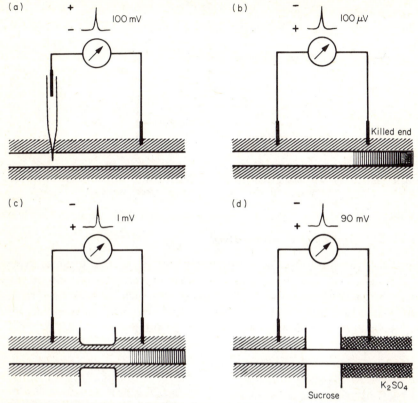

Fig. 3. The relation between intra- and extracellular recording. (a) A microelectrode records a 100 mV action potential from a large axon. (b) Extracellular recording, made monophasic by killing one end of the axon. The right-hand electrode behaves as an intracellular lead, so that the action potential is recorded with reversed polarity. (c) Obstruction of the extracellular space between the electrodes increases its resistance and thus gives a bigger signal. (d) Sucrose gap. The axonal membrane to the right of the gap is made physiologically inert by isotonic $K^+$, so that the right-hand electrode can function as an intracellular lead.

## C. Kostyuk's Method

An ingenious way of gaining electrical access to the inside of a cell membrane is shown in Fig. 4. The method is well adapted to fast voltage-clamping and in addition allows the composition of the intracellular medium to be altered at will (Kostyuk and Krishtal, 1977; Lee, Akaike and Brown, 1980). It has so far been applied mainly to large cells such as snail neurons and invertebrate oocytes.

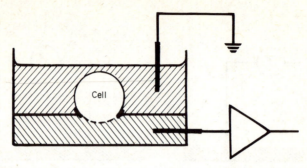

Fig. 4. Kostyuk's method of "intracellular" recording. A large cell is sealed over a hole, and the protruding membrane ruptured to provide electrical access to the intracellular compartment. No transmembrane electrode is used.

## D. *Voltage-sensitive Dyes*

Certain dyes have optical properties sensitive to the local electrical environment. Action potentials have been recorded optically from a variety of nerve and muscle cells stained with such dyes (for references see Morad and Salama, 1979). The method does not yet compete with intracellular recording, but may do so in the future.

# 2
# Making and Filling Microelectrodes

## I. Introduction

The user of microelectrodes spends a lot of time worrying about and experimenting with various methods of manufacture and filling. Consequently some parts of the process have become invested with ritualistic superstition. This is perhaps no bad thing, since it gives the electrophysiologist the feeling of belonging to a craft fraternity. However, in the account that follows I have tried to avoid propagating superstitious practices.

## II. Glass

Surprisingly, almost any glass capillary tubing of external diameter 1–3 mm can make good microelectrodes. Glasses as different as quartz and soda have been used successfully. Most people use borosilicate, with a softening temperature intermediate between those of quartz and soda glass. It is an advantage for making reproducible electrodes if the tubing is of uniform diameter. Some batches may need to be measured with a micrometer and graded according to size. The nominal diameter of the tubing selected does not matter much for most applications, although it is easier to make very fine-tipped pipettes from 1 mm tubing than from 3 mm tubing. Most microelectrode tubing has an inner diameter about half the external diameter. Special thin-walled tubing is now marketed by F. Haer & Co. (see below); it is said to produce ultrafine tips more easily than conventional tubing (Jacobson and Mealing, 1980).

Multi-barrelled electrodes are made from tubing which has been fused together by heating over a gas flame (Kennard, 1958; Curtis, 1964).

13

This handicraft is best learnt from someone who is already skilled at it. However, prefused tubing is available commercially for electrodes with 2, 3, 4, 5 and 7 barrels. An especially convenient way of making double-barrelled electrodes is to use "theta" ($\theta$) glass which has an internal partition dividing the bore into two halves.

Glass tubing and many other items of electrophysiological interest are available from several suppliers:

| Clark Electromedical Instruments, PO Box 8, Pangbourne, Reading RG8 7HU UK | F. Haer & Co., Brunswick, Maine 04011, USA | W-P Instruments, Inc., PO Box 3110, New Haven, Connecticut 06515, USA |
|---|---|---|

Some laboratories subject their glass to an elaborate sequence of chemical treatments. Like Thomas (1978), I have found this to be a waste of time, and use the glass straight from the package without washing or cleaning it.

## III. Pipette Pullers

Most of what needs to be said about pipette pullers has been said by Kennard (1958), Frank and Becker (1964), Curtis (1964) and Thomas (1978). If you want microelectrodes with tips greater than about 0.2 $\mu$m for impaling giant neurones or skeletal muscle, then any puller will do the job. You have only to learn the correct settings by trial and error, and get on with it. The difficulties arise when you have special requirements like ultrafine tips (less than 0.1 $\mu$m), multi-barrelled assemblies, or specified taper angle.

The *horizontal two-stage* puller based on the design of Alexander and Nastuk (1953) is the standard workhorse. Many laboratories have a "home-made" imitation. A weak sensing pull is applied to the glass while it is heating. At a predetermined elongation (2–5 mm) the pulling force is suddenly increased; it is this second, strong, pull that separates the tubing into two micropipettes. Frank and Becker (1964) give a wealth of information about the operation of this puller.

*Vertical two-stage* pullers allow the weak pull to be made by gravity (Winsbury, 1956). This type of puller is usually more massive than the horizontal type, has less acceleration during the strong pull, and perhaps for this reason cannot make such fine tips. But it can pull multi-barrelled assemblies from thick tubing.

The *horizontal one-stage puller* (Livingston and Duggar, 1934) is unique in not having a weak pull to sense the softening of the glass. Fierce heat is applied for a predetermined time from a U-shaped platinum foil heater, which needs to be almost white hot, and then the pipettes are pulled by a spring mechanism. This puller can make very fine tips but the shanks are unduly long.

To obtain ultra-fine tips without the long shank, additional refinements have been introduced. A precisely-timed jet of air or nitrogen directed at the heating coil increases the rate of cooling of the glass on each side of the tip region and gives shorter shanks without much effect on tip diameter (Chowdhury, 1969*a*; see discussion by Purves, 1980*a*). A fully engineered design has been given by Brown and Flaming (1977).

Ensor (1979) describes a puller whose moving parts are much lighter than usual. The resulting rapid acceleration, coupled with an adjustable cooling period between the weak and strong pulls, gives very fine tips with short shanks.

Heating elements may be made from a short coil of nichrome or kanthal resistance wire, or from a strip of nichrome, platinum or platinum/iridium 90%/10% alloy. Thomas (1978) lists some commercial suppliers. Strip elements may be shaped $\cup$ or $\Omega$ or $\circlearrowright$. None of these heating elements has a clear advantage. Platinum and platinum alloys can be run much hotter than nichrome, but the usual 80 $\mu$m foil is mechanically weak and easily bent out of shape. Evidently there is much scope for experiment with heating elements. A small close-fitting element can be run at a fairly low temperature (yellow or red-yellow heat); it does not heat a long length of glass and the resulting pipettes have fairly short shanks.

The design and operation of pipette pullers is almost entirely empirical (but see Chapter 3, Section III), the time-honoured maxim being "the hotter the heating element, the longer the shank and the finer the tip". To get reproducible results from any pipette puller, air currents should be excluded by a hinged transparent cover.

### IV. Filling Solutions

Salt bridges in electrochemical practice (Ives and Janz, 1961) are usually filled with concentrated solutions of KCl. The reasoning here is that $K^+$ and $Cl^-$ have similar diffusion coefficients and carry unit charge. Diffusion of KCl from a region of high concentration to a region of low concentration is therefore associated with only a small tendency towards charge separation, and the liquid junction potential is small. The choice of 3 M KCl to fill microelectrodes may be justified in the same way (Rae and Germer, 1974), but as shown in Chapter 3 Section VI the stratagem is not always successful since many microelectrodes exhibit a large tip potential. Thus there is no compelling reason to use 3 M KCl and it is readily found by experiment that almost any electrolyte solution will give usable microelectrodes (e.g. see Hill-Smith and Purves, 1978, for a recording made with 0.1 M cyclic AMP as the filling solution).

It is usual to make the filling solution fairly concentrated (1–3 M) with the aim of getting a high concentration of ions in the tip and hence a low electrical resistance. The diffusional mixing of inside and outside solutions within the microelectrode (Chapter 3, Section IV) means that the ion content near the tip is governed as much by the outside solution as by the inside, and so this stratagem too is not overly successful. Thus a 20-fold increase in concentration of the filling solution from 0.15 M to 3 M KCl reduces the resistance only sevenfold. The disadvantage of concentrated filling solutions is that they lead to a large diffusional leak out of the tip (Chapter 6, Section II). Leakage of $K^+$ is of little importance because cells already contain a lot of $K^+$ ions. But leakage of $Cl^-$ can be large enough to alter the $Cl^-$ equilibrium potential and reverse the sign of inhibitory post-synaptic potentials (see Thomas, 1978). Unwanted diffusional leakage of a particular ion can be reduced by using micropipettes of smaller tip diameter or by filling them with more dilute solution; it can be eliminated by choosing a filling solution which does not contain that ion. The usual alternatives to KCl are 1–2 M potassium acetate or potassium citrate (Rosenberg, 1973). It is said that if microelectrodes are to be stored for more than a few hours the pH of the filling solution should be adjusted to be less than 7 (add a few drops of 0.1 M hydrochloric acid) to prevent alkaline attack of the glass. The tip potential of KCl-filled microelectrodes may be made smaller if the pH is dropped to about 4 (Chapter 3, Section VI). If contact with the filling solution in the stem is made with a silver/silver chloride electrode, the $Cl^-$ concentration in the solution should be at least 5 mM (Thomas, 1978) to ensure electrochemical reversibility and stability of the half-cell potential.

Some people go to fanatical lengths to ensure purity and cleanness of their filling solutions. Certain filling methods (Section V in this chapter) are very sensitive to the presence of suspended particles. Others, like the glass fibre method, are not. Hospital pharmacies can be induced to supply large volumes of filling solutions, filtered through pores of 0.22 $\mu$m. This is clean enough for most purposes. Expensive filling solutions which are available in small volumes only (e.g. drugs for ionophoresis) can be cleaned without waste by centrifuging them at about 3000g for 10 min in a small tube.

### V.  Filling Methods

Why is it difficult to fill a micropipette? Most workers when asked this question assume a hang-dog expression and mutter something about "surface tension in a very small capillary". In fact the surface energies are in your favour when you are trying to introduce an aqueous solution into a glass tube. Water would rather be in contact with glass than with air, as illustrated by the well-known phenomenon of capillary rise. The real culprit is viscosity.

According to the Poiseuille – Hagen relation, viscous resistance to flow in a cylindrical tube is inversely proportional to the fourth power of the radius. In a sufficiently long conical tube, viscous resistance is inversely proportional only to the third power of the tip radius, but even so the flow through a fine-tipped pipette is clearly going to be very slow indeed. Filling methods in which a substantial volume of water has to enter via the tip, or in which a substantial volume of air has to leave via the tip, are likely to be successful only for rather coarse pipettes.

I have assembled here what I hope is a complete list of the astonishingly large number of different filling methods. Some appear with numerous small variations, but I have not detailed all of them. The principal advantages and disadvantages of each method are given individually. In most applications the method of choice (for novice and old hand alike) is the glass fibre method (Section M, below).

### A.  Direct Through Tip

If a micropipette tip is lowered into the filling solution the interface between air and water (Fig. 5(a)) assumes a curved shape with a radius of curvature $r$ equal to the internal radius of the tip. The pressure drop $\Delta P$ across the interface is given by the Laplace relation $\Delta P = 2T/r$, where $T$ is surface tension. The direction of the pressure difference is such as to favour hydrodynamic flow into the pipette, and filling proceeds rapidly at first. Three influences immediately begin to slow the rate of advance. (1) The radius of curvature increases since the pipette is tapered. This decreases the driving pressure. (2) Viscous resistance increases as the column of liquid becomes longer. (3) As the tapering column lengthens, a larger volume of liquid is needed to produce unit increase in length. The combination of these three effects brings about a disastrous fall in the rate of filling.

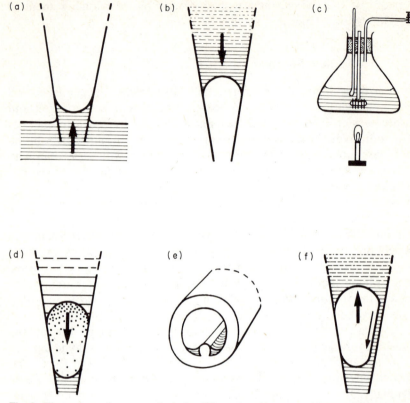

Fig. 5. Illustrations of some methods for filling microelectrodes. (a) Direct through the tip. Note the meniscus. (b) Direct through stem. The meniscus curves the opposite way to that in (a). (c) Apparatus for boiling method. (d) Distillation method. Water molecules diffuse towards the tip. (e) Glass fibre method. Hydraulic conduits are shown on each side of the fibre. (f) Glass fibre method, showing the mechanism for bubble transport.

In a sufficiently coarse micropipette whose filled resistance would be less than about 10 MΩ the interface may advance far enough over a few hours or days to become accessible via the stem and shank. Filling solution is then injected into the shank with a fine syringe needle and any remaining bubble dislodged with an animal whisker, a fine wire or a glass fibre (Nastuk, 1953*a*; Caldwell and Downing, 1955; Kurella, 1958; Lassen and Sten-Knudsen, 1968; Krischer, 1969*b*; Okada and Inouye, 1975).

*Advantages*:    cold-filling,    need    not    waste    filling    solution.
*Disadvantages*:    slow,    unreliable,    works    only    for    low-resistance microelectrodes, filling solution must be free from suspended particles.

### B. Direct Through Stem

This method illustrates the fact that it is easier to pass air than water through a narrow tube, because air has a low viscosity and is compressible. Filling solution is injected as far down the shank as possible with a microsyringe needle or fine cannula (Robinson and Scott, 1973) or pushed along with a probe (Mullins and Noda, 1963). The pressure drop across the meniscus near the tip (Fig. 5(b)) is such as to advance the filling solution at a rate which is initially small but increases as the tip is approached. Robinson and Scott (1973) measured the rate of advance under a microscope and used this value in an ingenious calculation of the internal tip diameter of their microelectrodes. Their method of calculation probably overestimates the diameter since they neglected to correct for the mean free path of gas molecules, which at atmospheric pressure is of the order (0.1 $\mu$m) of the tip dimensions.

*Advantages*: cold-filling, does not waste filling solution; applicable to fairly small-tipped pipettes. *Disadvantages*: needs manual dexterity, filling time unpredictable.

### C. Boiling

An understanding of the physical principles will enable this method to be used efficiently. The idea is firstly to replace the air content of the pipettes with water vapour via the stem, and secondly to allow the vapour to condense so that the filling solution can enter. In practice the two steps are not completely distinct. Replacement of air by water vapour takes place by diffusion. In aqueous solution diffusional equilibration along the whole length of a pipette would take many hours or days, but the coefficient of interdiffusion of air and water vapour $(0.23 \text{ cm}^2 \text{ s}^{-1})$ is so much greater than the diffusion coefficient of ordinary solutes $(0.5 \times 10^{-5} - 5 \times 10^{-5} \text{ cm}^2 \text{ s}^{-1})$ that a few minutes may suffice. To get the air out quickly the open ends of the stems should be exposed to water vapour, and the stems should not be blocked by liquid. The primitive boiling method (Ling and Gerard, 1949), in which the pipettes are suspended in filling solution and simply boiled vigorously over a gas flame, has little to recommend it since neither of the above conditions is met.

A much better way (Winsbury, 1956) is as follows. Suspend the pipettes tip down in filling solution in a glass flask whose bung is fitted with a thermometer and suction tube (Fig. 5(c)). Heat the solution to 70 or 80 °C. *Turn off* the heat and apply suction from a water pump, regulating it so that a strong stream of bubbles arises at the open ends of the pipette stems. There is no point in having explosive boiling throughout the flask, nor is it sensible to continue the heat application. Continued heating only leads to boiling at the bottom of the flask. Filling should be complete when the temperature has fallen to 45 or 50 °C.

*Advantages*: works for all microelectrodes, no individual treatment needed, large batches can be processed simultaneously. *Disadvantages*: high temperature damages tips, large tip potentials, wastes filling solution, unstable compounds cannot be used.

## D. Vacuum

The air is removed from microelectrodes by placing them in a container connected to a vacuum pump. After the vacuum is established, filling solution is introduced so that it covers the pipettes. Filling proceeds spontaneously. Atmospheric pressure is restored and the electrodes kept for a day or so before use (Schanne, Lavallée, Laprade and Gagné, 1968).

*Advantages*: as for boiling method, cold-filling. *Disadvantages*: needs vacuum technology, wastes filling solution.

## E. Alcohol

The idea behind this method is that alcohol can be made to boil at room temperature under reduced pressure. Some of the disadvantages of the ordinary boiling method are overcome. The sequence of events is that air is replaced by alcohol vapour, the vapour is condensed, distilled water replaces most of the alcohol, and finally the filling solution replaces water.

The original version (Tasaki, Polley and Orrego, 1954) used ethanol. The method I used routinely for several years was passed on to me by a colleague and I have not found a published account. Micropipettes are attached to a convenient holder and immersed tip down in methanol. Suction is applied from a water pump and controlled to produce boiling as in method C. After 5 min the pipettes are removed and immersed in distilled water for about half an hour. Next they are removed from the water and the stems are emptied with a 28 or 30 G needle introduced as far as the shoulders. The filling solution is injected in the same way and the microelectrodes are left for 12–24 h in a sealed container with a few drops of water in it to maintain high humidity. If the filling solution is cheap and available in quantity, some time can be saved in the previous step if an excess is injected into the stems, washing away the distilled water content.

An important part of the procedure is the time for diffusional equilibration; electrolyte in the filling solution has to reach the tips in appreciable concentration or else the electrical resistance of the microelectrodes will be very high. Calculations suggest that 95% equilibration ought to be reached in 1–3 days, depending mainly on the shank length. In practice most electrodes are usable after storage overnight, although the resistance continues to fall slowly over several days (Okada and Inouye, 1975).

*Advantages*: cold-filling, works for all micropipettes (even those with sealed tips), does not waste filling solution. *Disadvantages*: methanol is toxic and should be used in a fume cupboard, rather slow, moderately fiddly, large tip potentials.

### F. Pre-filled

Kao (1954) described an improbable-sounding method in which the filling solution is introduced into the glass tubing before the pipettes are pulled. I have made an attempt at this method but never got it to work. An improved version was used with evident success by Crain (1956).

*Advantages*: rapid. *Disadvantages*: clumsy modification to pipette puller needed.

### G. Centrifugation

This is really a variant of method B. The driving force for filling is increased by centrifugation. Coarse-tipped (2 $\mu$m) pipettes for ionophoresis in the central nervous system can be filled by centrifugation at 1600**g** for 5–10 min (Curtis, 1964; see also Zimmerer, 1973). Centrifugation at 19 000**g** is claimed to fill a 0.2 $\mu$m tip in 3 min (Cerf and Cerf, 1974). Henderson (1967) used 30 000**g** for 15 min.

*Advantages*: cold-filling, does not waste filling solution. *Disadvantages*: filling time depends on tip diameter.

### H. Pressure

Another variant of method B. The driving force is increased by pressure applied to the open end of the stem. Applicable only to very coarse pipettes.

### I. Distillation

An interesting family of filling methods relies on a difference of vapour pressure to transport water along the shank. Filling proceeds by evaporation from a liquid phase in the stem, diffusion in the vapour phase in the shank, and condensation in the terminal shank near the tip. In the original method of Caldwell and Downing (1955) (see also Byzov and Chernyshov, 1961; Chowdhury, 1969b), the tip of the pipette is immersed in filling solution which enters the tip for a small way as described in Section A above. Distilled water is then introduced into the stem. The vapour pressure in equilibrium with the filling solution at the tip is depressed according to Raoult's law, and the resulting difference in vapour pressures between the stem and tip provides the driving force for diffusional transport (Fig. 5(d)). Distillation is allowed to continue until the air bubble in the shank can be dislodged (see method A).

As we saw in method C, diffusion in the vapour phase should be rapid, and it is necessary to explain why the unmodified distillation method is in fact very slow, taking several days. Obviously this is because the solution in the tip becomes diluted by condensed water vapour until its vapour pressure approaches that of pure water. A subsidiary factor is that heat must be supplied to evaporate water from the stems. This suggests a simple improvement: if the water in the stem is heated, its vapour pressure will be increased. Larger gradients can be obtained in this way than those due to Raoult's law. The author's method is to put diluted Indian ink in the stems instead of water. The shank is immersed in a large volume of filling solution or water and the pipettes are exposed to a source of radiant heat which warms the stems differentially. Distillation time is reduced to an hour or less. Particles in the ink do not block the tips as they are not transported by distillation, and are flushed out of the stems later.

In all distillation methods the shank of the microelectrodes contains nearly pure water. The final stages of filling must include a step in which the stem is flushed out with filling solution and bubbles dislodged, and a period for diffusional equilibration as in method E.

*Advantages*: cold-filling, reliable, need not waste filling solution. *Disadvantages*: slow.

### J. Local Heating

Strictly this is a variant of the distillation method. Strong radiant heat is applied near the shoulder of individual pipettes whose stems contain water (Henderson, 1967). Condensation can be induced to occur at the tip without preliminary filling by capillarity (Zeuthen, 1971; Vyskočil and Melichar, 1972). The subsequent procedure follows that of method I.

*Advantages*: does not waste filling solution, semi-cold-filling, faster than distillation method. *Disadvantages*: fiddly manipulation of individual pipettes, equilibration time needed, semi-hot-filling.

### K. Vibration

Chowdhury (1969*b*) describes a technique whereby pipettes are vibrated axially with a piezoelectric device at high frequency. Chowdhury also indicates other, perhaps more interesting, uses for the vibrator in microinjection and impalement of cells.

### L. Mobile Filament

Coarse-tipped pipettes can be filled by injecting a small drop of filling solution into the shank and then inserting a *very* fine wire or glass filament through the drop towards the tip. Alternate withdrawal and insertion allows solution to reach the tip (Nastuk, 1953*a*; Ito, Kostyuk and Oshima, 1962).

### M. Glass Fibre

This brilliantly successful method was invented by Tasaki, Tsukuhara, Ito, Wayner and Wu (1968) who chose to report it in the journal *Physiology and Behaviour*. They inserted 5–30 fibres of about 20 $\mu$m diameter into each electrode blank and then pulled pipettes in the usual way. Filling was accomplished within a few seconds by injection of the solution at the shoulder. Subsequently it was found that even a single fibre suffices, provided it is not of too small a diameter (e.g. Thomas, 1972). During the pulling of a micropipette the introduced fibre or fibres fuse with the wall of the shank and tip (see the scanning micrograph of Brown and Flaming, 1974). It is important that the softening temperature and coefficient of thermal expansion of the fibre match those of the tubing. If you want to make your own you can easily hand-draw 100 $\mu$m fibres from the same tubing which the pipettes will be pulled from. Commercial tubing has the fibre fused to the wall to stop it falling out and is much more convenient to handle.

How does it work? On each side of the fibre where it is fused to the inside bore of the tubing there is a hydraulic conduit (Fig. 5(e)) along which filling solution advances by capillarity. Most of the cross-sectional area of the pipette is available for simultaneous transport of air in the reverse direction. With continued injection, a bubble forms in the shank (Fig. 5(f)). A conventional pipette would be unable to eject the bubble and would be useless. A fibre-containing pipette can usually eject the bubble so long as the shank is tapered, for in this case the radii of curvature of the two menisci are different and there is a favourable pressure gradient. The fibre again provides hydraulic conduits to pass the filling solution towards the tip. Actually, bubbles in the shank are nearly always harmless anyway, because there is electrical continuity on each side of the fibre. "Theta" glass (Section II in this chapter) can be filled in the same way as fibre-containing pipettes and for the same reasons.

*Advantages*: rapid, cold-filling, insensitive to particulate contamination, pipettes can be filled as needed during experiment, multi-barrelled assemblies can be filled with different solutions. *Disadvantages*: none. The expense of commercial fibre-containing tubing is sometimes supposed to be an important disadvantage, but the savings in time alone repay the cost many times over. In any case the glass for a day's experimentation costs about the same as a bar of soap.

One advantage claimed by Tasaki *et al.* (1968) was that the microelectrodes had a much lower resistance than those filled by other methods. I do not think that the claim can be upheld, although improperly filled electrodes may show higher resistances than they ought to (see discussion of method E above).


## VI. Bending, Breaking and Bevelling

Occasionally it is necessary to bend the shanks of microelectrodes to prevent them from obstructing the objective lens of a microscope. After much practice this can be done merely by holding them close to an electrically heated wire. An elegant refinement is detailed by Hudspeth and Corey (1978).

A microelectrode is normally ready for use when it has been pulled and filled. Some workers find the electrical properties to be much improved by partial breakage of the tip. This is especially true for ionophoretic pipettes which do not have to penetrate cells. There are many reports of methods for "controlled breakage" (Frank and Becker, 1964; Curtis, 1964), mostly superseded now by the bevelling techniques.

Bevelling is the removal of part of the tip of a micropipette so that it comes to resemble the tip of a hypodermic needle instead of being square-cut at the end. Having wasted a lot of time several years ago attempting to bevel microelectrodes before the technology had been properly worked out, I am a confirmed non-beveller. But bevelling appears to offer significant advantages (Isenberg, 1979): (1) The resistance of a bevelled microelectrode may be less than half its unbevelled value. The numerous benefits of low resistance are mentioned in Chapter 3, Section IV. (2) The very sharp tip of a bevelled microelectrode penetrates tissues and cell membranes more easily and produces less mechanical distortion. (3) Bevelled tips are less liable to blockage. The disadvantages are: the expensive apparatus needed for the usual grinding method, and the increased time required to prepare microelectrodes. If one habitually uses 30 or 40 microelectrodes in a day, the extra 5–10 min per electrode will not leave much time for experimentation.

Since I do not bevel my electrodes I cannot comment usefully on the details of the method. A list of references to the grinding technique will be given; anyone contemplating bevelling will need to read these carefully: Barrett and Graubard (1970); Shaw and Lee (1973); Brown and Flaming (1974, 1975, 1977, 1979); Kripke and Ogden (1974); Proenza and Morton (1975); Tauchi and Kikuchi (1977); Baldwin (1980a). An extremely interesting alternative method, in which "a jet of grinding solution is simply squirted against the electrode tip" has been described by Ogden, Citron and Pierantoni (1978). Related methods, which also require very little apparatus, are discussed by Lederer, Spindler and Eisner (1979) and Corson, Goodman and Fein (1979).

# 3
# The Physics of Microelectrodes

## I. Introduction

The accuracy with which measurements can be made using microelectrodes is in many instances limited by the physical properties of the electrodes themselves. Some understanding of the physics should help the user to recognize the limitations of microelectrode techniques, and to plan the details of his experiments so that the limitations are avoided. However, much good microelectrode work is done by people with little or no interest in physics; the contents of this chapter may be regarded more as reference material than as essential knowledge.

There does not appear to be any unified or general review of the physics of microelectrodes. The book "Glass Microelectrodes" edited by Lavallée, Schanne and Hébert (1969) comes closest, although its individual chapters are written by different authors and deal with only a few specific topics, often in an idiosyncratic way. Other accounts are given by Rush, Lepeschkin and Brooks (1968), Schanne et al. (1968) and Firth and DeFelice (1971). Microelectrode noise is discussed briefly in Chapter 4, Section VI. Diffusion and ionophoresis are considered in Chapter 6, Section II.

## II. Size and Shape

The naked eye appearance, and probably the light microscopic appearance of microelectrodes will be familiar to all readers. It is an unfortunate fact that the light microscope fails to give much useful information about the tip geometry and diameter, except perhaps for the coarse (2 $\mu$m) pipettes employed for ionophoresis in the central nervous system. Electron microscopes can give excellent views of even the smallest tips. It is interesting if not otherwise rewarding to look at the pictures published by various authors: transmission electron microscopy (Alexander and Nastuk, 1953; Amatniek, 1958; Byzov and Chernyshov, 1961; Bils and Lavallée, 1964; Frank and Becker, 1964; Forman and Cruce, 1972); scanning electron microscopy (Brown and Flaming, 1974, 1977, 1979; Fry, 1975; Ogden *et al.*, 1978). The method of Fry (1975) and Baldwin (1980*b*) is notable in that it does not damage the tips; the microelectrodes can be filled and used after viewing.

Tip diameters down to 0.048 $\mu$m have been reported by Brown and Flaming (1977) for borosilicate pipettes pulled on their high-performance puller. Fry (1975) shows a soda glass pipette which tapers to a point, no hole being visible at a magnification of $\times 40\,000$. The profile of micropipettes near the tip depends on the type of glass used, and there may be quite large deviations from conical form. In a series of observations with four different glasses, each was found to produce a characteristic and instantly recognizable tip profile (D. M. Fry and R. D. Purves, unpublished observations). The relation between external tip diameter and electrical resistance (Section IV in this chapter) is affected by the angle of taper ($2\theta$ in Fig. 6) and by the ratio external diameter/internal diameter. Values of $2\theta$ are in the range $2$–$15°$ and usually $4$–$10°$, although if the tip is not conical the measurement has no precise meaning. The angle of taper is an important determinant of the speed of ionophoretic release (Chapter 6, Section II). It is widely believed that the external/internal diameter ratio at the tip is very close to that of the original tubing. In fact with most types of glass the wall tends to be relatively thinner at the tip (Fry, 1975; Ujec, Vít, Vyskočil and Králík, 1973).

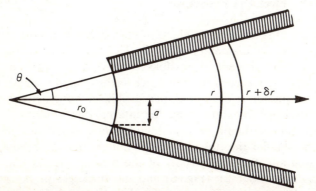

Fig. 6. Conical tip geometry assumed for theoretical treatment of microelectrode resistance. The internal tip diameter is $2a \approx 2r_0\theta$ and the included angle is $2\theta$.

### III. Mechanics

The mechanical properties of the fully formed pipette require little comment. Very long pipettes made on a one-stage puller are extremely flexible. Sometimes this is a disadvantage since it appears to magnify vibrations. Sometimes, though, it is an advantage in that it may allow an impaled muscle cell to move laterally without dislodging the impalement.

An interesting and useful feature of ultra-fine glass fibres, including micropipette tips, is that they are considerably stronger than macroscopic pieces of glass. The tensile strength of most materials is limited by surface irregularities ("Griffith cracks") but in freshly drawn fibres the irregularities are smeared out or obliterated by the drawing process, and so the tensile strength approaches the theoretical maximum (see Gordon, 1976). This effect probably increases the strength of micropipette tips by at least an order of magnitude. Without it, microelectrode breakage would be even more common and irritating than it actually is.

The high tensile strength of small fibres is also of importance in the pulling of micropipettes. The pulling process is a non-uniform viscous deformation under uniaxial stress. Separation of the tubing into two pipettes occurs when the tensile stress exceeds the strength of the glass. Most of the microelectrode profiles shown in this book were drawn by an X–Y plotter under computer control, using an equation derived by Purves (1980a). Interested readers are referred to this study for further details of the physics of pipette pulling.

### IV. Electrical Resistance

In the absence of some form of electron microscopy the most convenient way to estimate tip diameters of microelectrodes is to measure their resistance. It is not in general possible to arrive at an accurate value of the internal tip diameter from resistance measurements alone, and the external tip diameter is of course inaccessible, although for most types of glass tubing it will be about twice the internal diameter.

The usual rule of thumb is: other things being equal, the higher the resistance the smaller the tip. The "other things" are the conductivity of the filling solution, the conductivity of the medium outside the tip, the included angle of taper $2\theta$ (Fig. 6) and the thickness of the glass wall near the tip. In practice the rule of thumb works well. One soon learns the minimum resistance a microelectrode can have and still impale cells without too much damage (e.g. 5–15 MΩ for skeletal muscle, 25–50 MΩ for smooth muscle).

The resistance of the microelectrode is also important for completely different reasons. Too high a resistance will have adverse or even disastrous effects on tip potential (Section VI in this chapter), current passing ability (Section VII in this chapter and Chapter 5, Section V), noise (Chapter 4, Section VI), interference pick-up (Chapter 4, Section V), and input circuit time constant (Chapter 4, Section II). Specific means of dealing with these problems are discussed later in the appropriate sections, but it will be clear that from many points of view, the lower the resistance the better. The history of microelectrode technology can be regarded as a succession of attempts to minimize tip diameter and resistance simultaneously.

The serious practitioner, then, needs something more than a rule of thumb. We begin by deriving a very simple formula for microelectrode resistance to serve as a basis for more elaborate treatments. The internal bore of the pipette is taken as a cone of included angle $2\theta$, truncated at its tip by the sphere $r = r_0$ (Fig. 6). The angle $\theta$ in these and later formulae is understood to be measured in radians, although numerical values will be quoted in degrees. The internal tip diameter is $2a \approx 2r_0\theta$, and the conductivity $\sigma$ of the filling solution is supposed constant along the whole length of the pipette. To find the resistance we imagine a current $I$ to be flowing through the pipette, the flow being radially directed with respect to the point $r = 0$. The potential drop $\delta E$ across the shell between $r$ and $r + \delta r$ is proportional to $I$ and to $\delta r$, and inversely proportional to $\sigma$ and to the area of that part of the spherical surface of radius $r$ that is enclosed by the pipette walls:

$$\delta E = - \frac{I \, \delta r}{\psi r^2 \sigma} , \tag{1}$$

where $\psi \approx \pi\theta^2$ is the solid angle subtended by the pipette and the minus sign indicates that a positive potential gradient is associated with current flow in the negative $r$ direction (i.e. outward, towards the tip). Integrating between the tip $r = r_0$ and some large value of $r$ which for simplicity may be taken as infinite, we obtain

$$E = - \frac{I}{\pi\theta^2} \int_{r_0}^{\infty} \frac{dr}{\sigma r^2} \tag{2}$$

$$= - \frac{I}{\pi\theta^2 \sigma r_0}$$

and since $r_0 \approx a/\theta$, the resistance $R_{\mu E}$ is

$$R_{\mu E} = |E/I| = 1/\pi\theta\sigma a. \tag{3}$$

This result has been used fairly widely (Amatniek, 1958; Geddes, 1972) sometimes derived in a slightly different way so that $\tan \theta$ replaces $\theta$. For $\theta < 5°$ the two forms are virtually equivalent. The conductivity $\sigma$ of the commonest filling solution (3 M KCl) is about 26 S m$^{-1}$.

Lanthier and Schanne (1966) and Firth and DeFelice (1971) pointed out that equation (3) neglects diffusional mixing in the pipette tip of the filling solution and the electrolyte solution outside. If the two solutions have different conductivities, $\sigma$ in equation (1) is not a constant but is a function $\sigma = \sigma(r)$ of distance along the pipette. Under these conditions the integral of equation (2) cannot be used. The ion concentration at the extreme tip is in fact very nearly that of the external medium and independent of the filling solution, the asymmetry arising from the very small solid angle $\pi\theta^2$ subtended by the pipette, relative to the solid angle $4\pi - \pi\theta^2 \approx 4\pi$ outside.

Suppose the concentrations of the external and filling solutions to be $C_0^*$ and $C_0$ respectively, and suppose further that all ions present have the same diffusion coefficient $D$ and absolute charge number $|z|$. From the theory of diffusion (Crank, 1975) it can be shown that the total electrolyte concentration at any point in the pipette is

$$C = C_0\left(1 - \frac{r_0}{r}\right) + C_0^* \frac{r_0}{r}. \tag{4}$$

The conductivity of an ideal electrolyte solution is given by the Nernst – Einstein relation:

$$\sigma = 2z^2F^2DC/RT \tag{5}$$

where $F$ is the Faraday constant, $R$ is the gas constant and $T$ is the temperature. Substitution of equations (4) and (5) into equations (1) and (2) gives

$$E = -\frac{IRT}{2z^2F^2DC_0^*\pi\theta^2}\int_{r_0}^{\infty} \frac{dr}{wr^2 + rr_0(1 - w)} \tag{6}$$

and so

$$R_{\mu E} = \frac{RT \ln w}{2z^2F^2DC_0^*\pi\theta a(w - 1)}, \tag{7}$$

where $w = C_0/C_0^*$ and $\ln w$ is the natural logarithm of $w$. If we write $R^*$

Table 1

Calculated resistance of microelectrode as a function of internal tip diameter $2a$, taper angle $2\theta$, and concentration $C_0$ of the filling solution. The concentration $C_0^*$ of the external medium was taken as 150 mM, and all ions were assumed to have unit charge number and a diffusion coefficient $D$ of $1.7 \times 10^{-9}\,\text{m}^2\,\text{s}^{-1} = 1.7 \times 10^{-5}\,\text{cm}^2\,\text{s}^{-1}$.

| $2a/\mu\text{m}$ | $2\theta/°$ | $C_0/\text{M}$ | $R_{\mu\text{E}}/\text{M}\Omega$ |
|---|---|---|---|
| 0.5 | 8 | 3 | 1.5 |
| 0.1 | 6 | 3 | 10 |
| 0.02 | 4 | 3 | 76 |
| 0.5 | 8 | 1 | 3.2 |
| 0.1 | 6 | 1 | 22 |
| 0.02 | 4 | 1 | 160 |

for the resistance of the pipette for the special case $w = 1$,

$$R_{\mu\text{E}} = R^* \frac{\ln w}{w - 1}, \tag{8}$$

as shown by Firth and DeFelice (1971). The measured resistance of a pipette includes the "convergence" resistance to current flow in the external medium, but the required correction to equations (3) or (7) is negligibly small.

Some representative examples are given in Table 1, in which the concentration of the external medium has been taken as 150 mM, approximating that of physiological saline. These values underestimate the resistance somewhat, since equation (5) assumes the conductivity to be directly proportional to ion concentration, whereas the equivalent conductivity (Bockris and Reddy, 1970) of real electrolytes is a decreasing function of concentration. When other examples are calculated from equation (7) care should be taken to use a consistent set of units: in the Système International (SI), $a$ is expressed in metres, concentration in moles per cubic metre, and $\theta$ in radians.

The most important general lesson from equation (7) is that microelectrode resistance depends on both $C_0$ and $C_0^*$. The dependence on $C_0$ is therefore less than might have been expected. For example, a tenfold decrease in $C_0$ from 3 M to 300 mM increases the resistance by only 4.4 times.

## V. Capacitance

There are three chief components of the total shunt capacitance of the input circuit in microelectrode recording (Fig. 7): the input capacitance

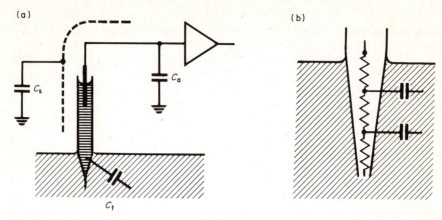

Fig. 7. Capacitance in the input circuit. (a) $C_a$, input capacitance of preamplifier; $C_s$, stray capacitance of connecting lead and stem of microelectrode; $C_t$, transmural capacitance between microelectrode and bathing solution. (b) The distributed nature of the transmural capacitance. $C_t$ may be regarded as the aggregate of a large number of small capacitances "mixed up" with the series resistances of the microelectrode. Only two components of $C_t$ are depicted.

$C_a$ of the preamplifier, the stray capacitance $C_s$ of the microelectrode's stem and the lead joining it to the preamplifier, and the transmural capacitance $C_t$ of that part of the microelectrode immersed in the bathing solution or tissue. Only the last-mentioned is discussed here. The total capacitance is important because, together with the microelectrode resistance, it forms a low-pass filter (Chapter 4, Sections II.C and II.D).

Textbooks of electronics or circuit theory give a formula for the capacitance $C'$ in picofarads per metre length of a coaxial cable:

$$C' = \frac{24.1\varepsilon_r}{\log_{10}(a_e/a_i)}, \tag{9}$$

where $\varepsilon_r$ is the relative permittivity ("dielectric constant") of the material separating the conductors and $a_e$, $a_i$ are the radii of the external and internal conductors. For glass, $\varepsilon_r$ is in the range 6–9, and the ratio of radii in the shank of a micropipette is in the range 1.4–2.2. Inserting these values we find $C'$ to be 0.4–1.5 pF per millimetre of immersed length. The overall transmural capacitance $C_t$ may therefore range from a fraction of a picofarad to 10 pF or more.

As shown in Fig. 7(b), $C_t$ is a "distributed" capacitance, the individual elements being connected by the series resistance of the microelectrode. Geddes (1972) explains this point fully. We may note in passing that distributed capacitance cannot be properly compensated by negative capacitance applied at the preamplifier (Chapter 5, Section V.F).

Note that $C_a$ and $C_s$ are generally capacitances to earth, but $C_t$ is a capacitance to the bathing fluid. The difference is important when pulsed voltages are applied to the bath (Chapter 4, Section II.D and Chapter 5, Section V.D).

## VI. Tip Potentials

The potential recorded by a microelectrode in the bathing solution or extracellular fluid is usually not zero. The total *offset potential* is considered to have three components, two of which are governed only by the nature and concentration of the various electrolyte solutions present, and the third of which depends in addition on the physical properties of the microelectrode tip. The two components of offset potential that are independent of the microelectrode are discussed elsewhere. They are the *liquid junction potential* (Chapter 1, Section II.H and Chapter 2, Section IV) and a potential arising from a possible dissimilarity between the indifferent electrode and the electrode which contacts the microelectrode's filling solution (Chapter 4, Section III).

The third component, or *tip potential* proper, in general is different for

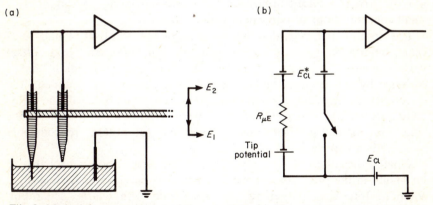

Fig. 8. Method for measurement of tip potentials. (a) The microelectrode under test is mounted alongside a broken-tipped microelectrode, shown on the right. The potential recorded when only the test electrode is immersed is $E_2$. When both electrodes are immersed the potential is $E_1$. Tip potential is $E_2 - E_1$. (b) Equivalent circuit of (a). $E_{Cl}$ is the half-cell potential established at the indifferent electrode, and $E_{Cl}^*$ is the corresponding potential at the electrodes that make contact with the microelectrode's filling solution. Lowering of the broken-tipped microelectrode into the bath is equivalent to closure of the switch.

different microelectrodes, but the differences are found experimentally to disappear if the tips are broken off. Most authors define tip potential as $E_2 - E_1$, where $E_2$ is the potential recorded by an intact microelectrode and $E_1$ is the potential after the tip has been broken. If the arrangement of Fig. 8 is used, tip potentials can be measured without actually breaking the microelectrode under test. Some authors follow a different usage, so that the term "tip potential" includes the liquid junction potential. A certain logic attends this usage, since the liquid junction potential assuredly results from events occurring in the tip. Moreover, a microelectrode that exhibits a large tip potential would not ordinarily be said to possess a liquid junction potential as well.

The behaviour of tip potentials has been studied by numerous authors (Adrian, 1956; Okada and Inouye, 1975; Wann and Goldsmith, 1972; Küchler, 1964; Schanne et al., 1968; Gotow, Ohba and Tomita, 1977; Riemer, Mayer and Ulbricht, 1974; Firth and DeFelice, 1971; Levine, 1966; Agin and Holtzman, 1966; Agin, 1969; Lavallée and Szabo, 1969).

In summary, the tip potential: (1) is of negative sign and of magnitude up to about $-70$ mV when the electrode is filled with 3 M KCl and tested in physiological saline; (2) is abolished when the inside and outside solutions are the same; (3) is larger in higher-resistance microelectrodes; (4) varies widely in magnitude between otherwise similar electrodes; (5) is often made smaller by use of filling solutions of low pH; (6) is reduced in size or even reversed in sign by addition of small concentrations of polyvalent cations like $Th^{4+}$.

The tip potential, regarded merely as an unwanted offset, would be of little importance since it could easily be compensated by an offset voltage control (Chapter 4, Section IV). The real problem is that *it changes* by an unpredictable amount when a cell is impaled, the direction of the change being such that the magnitude of the cell's resting potential is underestimated. Careful reading of Adrian's (1956) study is mandatory for anyone wishing to obtain reproducible measurements of resting potential.

What is the origin of the tip potential? The explanation is probably to be sought in the properties of the interface between glass and electrolyte solution. Interfaces in general are associated with charge separation (Shaw, 1970), and glass in contact with water acquires a layer of fixed negative charge. To balance this and restore overall electroneutrality there is a layer of positive charge in the adjacent solution. This arrangement, in its simplest conception, is known as the Helmholtz double layer; in more advanced treatments it is referred to as the Gouy – Chapman diffuse double layer. The positive layer is made up of mobile cations, anions being virtually excluded. Thus cations can diffuse along in the interfacial layer but anions cannot. The thickness of the double layer is of the order of $0.01\mu m$ on a plane surface, but possibly thicker inside a narrow tube like the tip of a microelectrode. In a sufficiently fine-tipped microelectrode the double layer may extend right across the lumen and form a kind of "cation plug" (Krischer, 1969a) preventing anion movement in or out of the tip. If the electrolyte concentrations inside and outside the microelectrode differ, a Nernst-like potential can develop whose sign and magnitude are governed by the ratio of cation concentrations.

The tip potential is usually smaller than the Nernst potential calculated from the concentration ratio, since the "cation plug" may not completely occlude the tip. The tip potential mechanism is therefore electrically shunted by the conductance of the bulk phase electrolyte. In a very coarse-tipped or broken electrode shunting becomes so great that the tip potential disappears, leaving only small and usually unimportant liquid junction potential.

Surface charge and other interfacial phenomena are difficult to study experimentally, since they are affected by trace contaminants, by the chemical and physical history of the interface, by electrolyte concentration and pH, and by many other influences, not all of which are easily controlled. Thus the tip potential, together with other manifestations of the double layer (Section VII in this chapter and Chapter 6, Section II.A), is characterized by great variability. The hopes expressed some years ago that treatment with polyvalent cations might reliably eliminate tip potentials do not seem to have eventuated. Selection of microelectrodes for small tip potential, according to Adrian's (1956) prescription, remains the chief defence against errors in the recording of resting potential, although some filling methods appear to produce larger tip potentials than others (Okada and Inoue, 1975; Plamondon, Gagné amd Poussart, 1979).

## VII. Non-linear Electrical Behaviour

Electrolytic conductors obey Ohm's law over the range of electric field strengths likely to be found within a microelectrode. One might expect then that the current – voltage relations of microelectrodes would be linear. Unfortunately they are not (Fig. 9), although quasi-linear behaviour is observed over a restricted range of applied voltages. Microelectrodes of low resistance may pass a few tens of nanoamperes while remaining in the quasi-linear region, whereas microelectrodes of

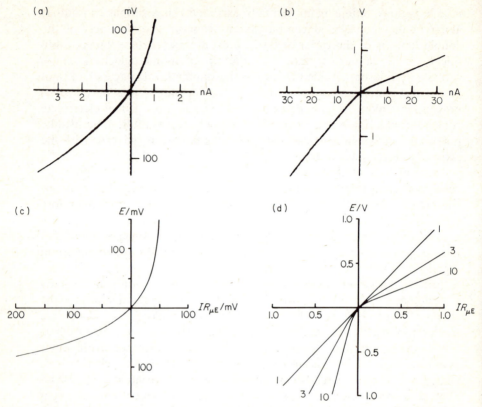

Fig. 9. Current – voltage relations of microelectrodes. In each case the upper right-hand quadrant represents the outward flow of current through the electrode. (a) Measured in an electrode of nominal resistance 50 MΩ, filled by the boiling method. (b) Measured in an electrode of nominal resistance 35 MΩ, filled by the glass fibre method. (c) Calculated for electrodes with a "cation plug" at the tip. (d) Calculated for electrodes with electro-osmotic bulk flow; numbers against curves are values of $w = C_0/C_0^*$. Details of the calculations are given in Appendix II. (a) and (c) show type I non-linearity. (b) and (d) show type II non-linearity.

high resistance (>50 MΩ) may be markedly non-linear for currents of 1 nA or less. The inconstancy of microelectrode resistances makes it difficult to pass precisely known currents by conventional means and provides (or ought to provide) strong motivation for the use of electronic current pumps (Chapter 5, Section II).

Applied voltages in excess of 1–10 V usually produce erratically varying current flow suggestive of intermittent blockage of the tip. Under these conditions it is hardly possible to speak of any single value of microelectrode resistance, and for most purposes the microelectrode is useless. This type of behaviour is perhaps due to boiling in the tip as a consequence of the very high power density (Rush *et al.*, 1968).

Most microelectrodes show reproducible non-linear current – voltage relations for applied voltages less than those which cause boiling. Experimentally, two kinds of non-linearity are seen. Qualitative descriptions of these, and of the probable underlying physics, follow. Attempts at a quantitative theory are to be found in Appendix II. In type I non-linearity (Fig. 9(a) and (c)) the resistance is higher for outward currents than for inward, and there is a strong suggestion of a definite maximum outward current which cannot be exceeded, no matter how large the applied voltage. The explanation of this behaviour follows immediately from the concept of a cation plug (Chapter 3, Section VI and see Krischer, 1969*a,b*). Outward currents drive cations through the plug into the medium outside. Anions are simultaneously transported backwards along the shank towards the stem. Thus a depleted region becomes established above the plug. The electrical resistance is high because there are few mobile charge carriers (ions) in this region. Conversely, inward currents increase the ion concentration above the plug and so the resistance falls.

Type II non-linearity shows the opposite dependence on current (Fig. 9(b) and (d)), outward currents being associated with a fall of resistance. Taylor (1953) suggested that electro-osmotic bulk flow (Chapter 6, Section II) might be the cause. As shown in Section IV of this chapter, the electrolyte concentration near the tip is intermediate between that of the filling and external solutions. Imposition of bulk flow through the pipette makes the concentration near the tip approach that of the filling solution (for outward flow) or external solution (for inward flow). If the conductivities of the two solutions differ, as is normally true, bulk flow leads to changes of resistance (Firth and DeFelice, 1971). The sign of the electro-osmotic flow is such that outward currents give outward flow and hence a fall of resistance.

In both kinds of non-linearity the change of resistance is secondary to a redistribution of ions. The change of resistance ought therefore to be time-dependent, a prediction borne out by experiment (Fig. 38). This fact is of great significance for current injection with simultaneous membrane potential measurement (Chapter 5).

The least satisfactory aspect of the above discussion is that we do not know why some electrodes show type I behaviour, and others type II. The latter appears to be favoured by the use of highly concentrated filling solutions, but the only systematic study along these lines is that of Krischer (1969*b*), who dealt exclusively with type I non-linearity.

Rational discussion of non-linear electrical properties of microelectrodes has not been made easier by the frequent quoting of a study (Rubio and Zubieta, 1961) in which the non-linearities were apparently of type I but whose explanation was sought in a type II mechanism.

# 4

# *Circuitry for Recording*

## I. Introduction

This chapter deals with the electrical aspects of intracellular recording. Most of it is written at an elementary level to help the novice; intrusion by mathematical formulae has been kept to what I hope is an acceptable minimum. Some familiarity with simple circuits is essential for proficiency in microelectrode use. Appendix I outlines suggested topics for study and introduces a number of technical terms.

The relative importance of the various subjects discussed in this chapter will of course depend on the application. If the aim is reproducible measurement of resting potentials, then the method of connection to the microelectrode, the nature of the indifferent electrode and an understanding of offset potentials will be of most interest (see also Chapter 1, Section II.H and Chapter 3, Section VI). If fast transient signals are to be recorded, stray capacitance and its compensation assume significance. And if very small responses are to be measured, interference and noise become all-important.

## II. Preamplifiers

An introductory account of the functions of a preamplifier was given in Chapter 1, Section II.C.

### A. Understanding the Specifications

Commercial preamplifiers come with a list of specifications many of which have been chosen, if not actually manipulated, to look as impressive as possible. Some specifications are unambiguous, can be checked by the user, and allow comparison between different amplifiers. Others, unfortunately the more important ones, depend on the fine adjustment of an internal or front panel control, or are quoted in different ways by different manufacturers; either way, comparison between amplifiers is difficult.

*Leakage current* or *bias current* is the current which flows into or out of the input terminal in the absence of applied voltages. Since this current traverses the microelectrode and impaled cell, it needs to be small enough to produce less than, say, 1 mV across the resistance $R_{\mu E}$ of the microelectrode. For resistances up to 100 MΩ, leakage current should be less than 10 pA. Leakage current is easily measured. Earth the input terminal and note the output voltage. Then earth the input via a resistor of high known value $R$ (e.g. 100 MΩ) and again note the output voltage. Leakage current is $\Delta E/AR$ where $\Delta E$ is the difference between the two readings and $A$ is the gain of the amplifier (in most preamplifiers $A = 1$).

The leakage current of the commonest input device, a field-effect transistor, doubles for each 10 °C rise in temperature. The specification of leakage current should state the temperature, although it rarely does.

Some amplifiers have provision for adjustment of leakage current. Specification is meaningless for these amplifiers since the value will depend mainly on how recently and carefully the adjustment was made. Resistors change their value slightly as they age and so the leakage current (which depends on very precise resistor matching in the internal circuitry) may change with time. Readjustment every few weeks or months is advisable.

*Input resistance* is another measure of how well the amplifier prevents current from flowing in the input circuit. It indicates how successfully the amplifier resists current flow when a voltage signal is applied. Normally one wants the input resistance to be at least 100 or 1000 times the microelectrode resistance. Modern input devices have little trouble in attaining $10^{11}$ or even $10^{12}$ Ω, adequate for normal microelectrodes but not for some ion-sensitive types (Thomas, 1978). If you use microelectrodes of 20 MΩ, an input resistance of $10^{11}$ Ω is just as good as an input resistance of $10^{13}$ Ω. The input resistance of preamplifiers equipped for current injection (Chapter 5, Section V) is governed by a resistor $R_s$ connected internally to the input, and by a preset adjustment of voltage gain. Like the leakage current above, input resistance of these amplifiers reflects how recently and accurately the adjustment was made. A bald specification is meaningless; this does not inhibit the manufacturers from quoting a value.

*Rise time and cut-off frequency* are expressions of the speed with which the amplifier can follow rapidly changing signals. All preamplifiers for intracellular recording have a bandwidth extending at the low end down to zero frequency (usually but illogically called d.c.; "d.v." = direct voltage would be better). Electrophysiologists are usually interested in "time domain responses", that is in waveforms as a function of time, and so the rise time $t_r$ is the most natural expression of an amplifier's response speed. Rise time $t_r$ is the time for the output to pass from 10% to 90% of the final value following the application of a step voltage change to the input (Fig. 10(a)). For faithful reproduction of fast signals like action potentials, $t_r$ should be less than a fifth of the signal's time to peak.

Cut-off frequency $f_{co}$ is the frequency at which the sine wave response of the amplifier falls to $1/\sqrt{2} \approx 0.7$ of its low frequency value (Fig. 10(b)). Cut-off frequency and rise time are related, accurately enough for practical needs, by the formula $t_r f_{co} = 0.35$ where $t_r$ is expressed in milliseconds and $f_{co}$ in kilohertz.

Rise time is usually quoted for zero source impedance, the voltage source being joined directly to the amplifier input. The value is usually impressively small, say $0.5\ \mu s$. This figure has little to do with the response speed when there is a high value resistor between source and input. Accordingly, most manufacturers also quote the rise time for a source resistance of 10 or 20 MΩ, obtained with the resistor connected very close to the input terminal to minimize stray capacitance. Values of

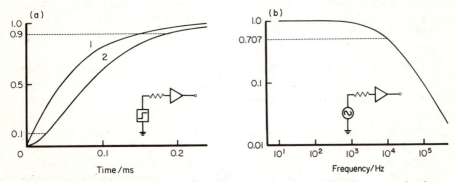

Fig. 10. Rise time and cut-off frequency. (a) Rise time is the time for the output signal to pass from 10% to 90% of its final value, following the application of a step voltage input signal (see Exercise 12 of Appendix I). The concept of rise time applies equally to an exponential waveform (trace 1) and a non-exponential waveform (trace 2). (b) Sine-wave frequency response corresponding to the step response (trace 1) of (a). Cut-off frequency, in this case 10 kHz, is the frequency at which the response falls to 0.707 of its low frequency value. Note the logarithmic scales.

5–15 $\mu$s are common for a 10 M$\Omega$ source and double that for a 20 M$\Omega$ source. Negative capacitance (Section II.D, in this chapter) will have been applied if available.

It would be useful if manufacturers could agree on a fixed method of specifying the rise time of a negative capacitance amplifier. To be realistic there should be some positive shunt capacitance for the compensation to "bite on". A suggested protocol is that the signal should be applied through two 10 M$\Omega$ resistors in series and shunted to earth by 10 pF. The uncompensated rise time would be about 440 $\mu$s. Negative capacitance would then be applied to give the shortest rise time without overshoot. Fast circuitry should be able to reduce the rise time to less than 30 $\mu$s.

*Absolute maximum input voltage* means exactly what it says. Even momentary application of voltages greater than this from a low impedance source will surely destroy a semiconductor input device. Vacuum tubes can tolerate much more abuse. For class or student use, semiconductor input stages should be protected with a resistor (47 k–470 k$\Omega$) connected in series as close as possible to the input device, the higher values giving protection up to perhaps 50–100 V. Some manufacturers reprehensibly fail to specify the maximum safe voltage.

*Working input voltage range* or *dynamic voltage range* is the range of input voltages over which the amplifier transmits signals to the output. If the working range is less than ±150 mV it may on occasion be exceeded by voltage offsets arising from electrochemical asymmetries in the input circuit (Section III of this chapter). In preamplifiers with current injection, the working voltage range needs to be large so that the amplifier can force large currents through high resistance electrodes. Values in excess of ±1 or 2 V are acceptable.

*Offset adjustment range* is not an important specification. In some amplifiers, the offset control effectively extends the working input voltage range; this is highly desirable if the latter is small. In others the offset control merely shifts the output voltage. There are two conflicting requirements in the design of offset controls. One is that the range should be fairly large (at least 150 or 200 mV) and the other is that it should be possible to make small adjustments of the order of 1 mV or less. The two requirements cannot be met by a single one-turn control. Either a multi-turn control or separate COARSE and FINE controls are needed.

*Drift* and *stability* are expressed in so many different ways that it is hardly possible to compare amplifiers. Common values are 50 $\mu$V per hour, 200 $\mu$V per day, or 150 $\mu$V per kelvin, none of which give cause for alarm. A thermal drift greater than about 200 $\mu$V per kelvin should be regarded with suspicion, and suggests that the amplifier may not be quite stable enough for accurate monitoring of resting potentials.

### B. What's Inside

The requirements of microelectrode preamplifiers have always been near the limits of contemporary technology. Developments over the last 25 years have led to improvements in drift, gain accuracy, and ease of design, but have not otherwise revolutionized performance.

Some manufacturers attempt to keep the details of circuit design of their preamplifiers secret; others are more forthcoming, although few publish complete details of the probe circuit. Modern preamplifiers almost without exception have a field effect transistor (FET) as the input device and integrated operational amplifiers in the remaining circuitry. FETs have characteristics somewhat similar to those of thermionic valves (high input resistance, small leakage current) but exhibit a large thermal drift (2.2 mV per kelvin). They have therefore to be used in pairs, combined in such a way that the individual drifts cancel. Best matching is obtained from two FETs fabricated together on the same silicon wafer. The "dual FET" in one or another form is the universal input stage.

Popular input circuits are shown in Fig. 11. The circuit of Fig. 11(c) perhaps represents the ultimate in simplicity, since from the user's stand-point it has only one component. With the addition of two 9 V batteries for the power supply, a chart recorder, and some means of adjusting offset voltages, this could form the entire electronic apparatus needed for studies of steady or not-too-rapidly varying intracellular potentials. Suitable FET operational amplifiers include, in increasing order of price and low noise performance, the R.C.A. CA 3140, the Teledyne-Philbrick 1421 and the Burr-Brown 3523. The cost and complexity of commercial preamplifiers comes from extra features like negative capacitance, offset voltage control, microelectrode resistance measurement, current injection, signal filters, and so on. It is quite possible to build preamplifiers incorporating some or all of the features and having a performance not markedly inferior to the commercial designs (Appendices III and IV; Colburn and Schwartz, 1972; Unwin and Moreton, 1974; M. V. Thomas, 1977; Muijser, 1979).

This is perhaps the place to comment on "home-made" versus commercial electronic apparatus. In the small volume market for electrophysiological apparatus, the price of commercial units is likely to be about five to eight times the cost of the components. Some workers,

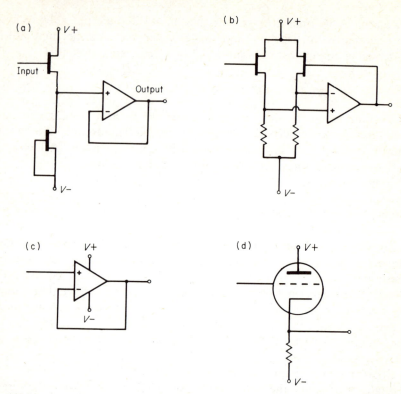

Fig. 11. Input circuits of microelectrode preamplifiers. (a) and (b) Circuits using a discrete dual FET and an operational amplifier. (c) Circuit using a FET-operational amplifier connected as a voltage follower (see Exercise 17 of Appendix I). (d) Cathode follower using a triode vacuum valve. This type of circuit is nearly obsolete.

especially those with restricted research budgets, feel that there is a strong economic incentive for making their own instruments or having them made in the departmental electronics workshop. "Home-made" instruments are not only cheaper than commercial equivalents, they may on occasion be better since the design can be tailored to suit local requirements. This presupposes sufficient technical expertise to bring the design and construction to a successful conclusion. Lastly, there is a good deal of satisfaction to be obtained from using apparatus built to one's own specification.

The other side of the argument is that commercial units are far more likely to work when first switched on. Also they come with a guarantee, a more or less well-written instruction manual, and an elegantly finished front panel.

### C. The Driven Shield

A recurring theme in microelectrode technology is the *low-pass filter* (Section VIII in this chapter) formed by the resistance $R_{\mu E}$ of the microelectrode and the total stray capacitance $C_{tot}$ of the input circuit (Figs 7 and 12). The effect of this low-pass filter on the response speed of the system is expressed as the time constant $R_{\mu E}C_{tot}$ (see Exercises 11–13 of Appendix I), the rise time $t_r \approx 2.2\ R_{\mu E}C_{tot}$, or the cut-off

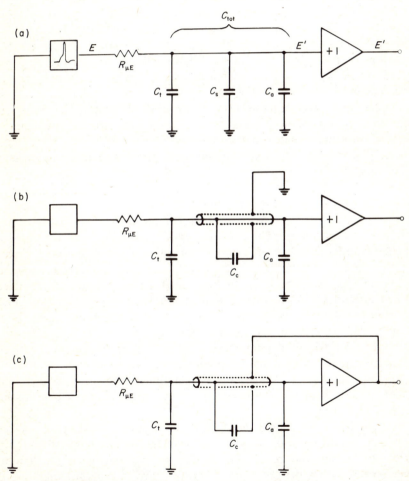

Fig. 12. Development of the driven shield. (a) The three components of the total capacitance $C_{tot}$ of the input circuit. (b) An earthed shield eliminates $C_s$ but replaces it by a much larger capacitance $C_c$. (c) The shield is driven at unity gain, i.e. it is kept at the same potential as the input signal.

frequency $f_{co} = 1/2\pi R_{\mu E}C_{tot}$ (see Section II.A in this chapter). As noted earlier, for faithful reproduction of fast transients the rise time should be at most a fifth the time to peak of the signal. An alternative way of looking at the harmful effects of $C_{tot}$ is as follows. In Fig. 12(a) the true signal voltage is $E$ and the filtered signal seen by the preamplifier is $E'$. Whenever $E$ and therefore $E'$ are varying in time, a current of magnitude $C_{tot}dE'/dt$ must flow through $C_{tot}$ to earth (this is an inescapable consequence of the definition of capacitance). The only place the current can come from is through the microelectrode resistance $R_{\mu E}$; by Ohm's law there must be a voltage drop $R_{\mu E}C_{tot}dE'/dt$. This represents the difference between the true signal $E$ and the distorted version $E'$. Clearly the distortion is worse the faster the variation of the signal. The general approaches to increasing response speed are obvious: use low resistance microelectrodes if possible, don't immerse the tip deeply in the bathing fluid or tissue, don't use a long wire to join the microelectrode to the preamplifier. The last recommendation when pushed to the limit leads to the now almost universal use of a small probe containing the preamplifier's input stage. The microelectrode plugs straight into the probe and there is virtually no connecting wire.

The driven shield (Nastuk and Hodgkin, 1950) is an alternative or additional way of minimizing the harm done by the stray capacitance $C_s$. It does not compensate for the other components (Fig. 12(a)) of the total shunt capacitance. The principle employed is that when a signal is applied simultaneously to the two terminals of a capacitor, no charging or discharging current flows. The development of the driven shield is shown in Fig. 12(b) and (c). In Fig. 12(b) an earthed shield has been placed around the input wire. This is evidently a retrograde step, since the core-to-screen capacitance $C_c$ is likely to be many times bigger than $C_s$ was. The only advantage of this connection is that it screens against alternating electric field interference. In Fig. 12(c) the shield has been connected to the output of the preamplifier, whose gain $A$ is close to 1.00. Thus the signal voltage appears simultaneously on both core and screen. The effective core-to-screen capacitance can be shown to be $C_c(1 - A) \approx 0$.

Use of a driven shield must not be regarded as a panacea for problems caused by poor physical layout of the input circuit. It is always better to reduce the real value of the stray capacitance $C_s$ than to overcome it with a driven shield. There are two reasons for this: one is related to noise (Section VI in this chapter), and the other is that, owing to the finite bandwidth of the amplifier, $A$ becomes less than 1.00 at high frequencies so that the shield potential does not "track" the signal.

It is therefore desirable to reduce the core-to-screen capacitance of the cable itself, either by keeping it short, or by using special low capacitance cable with a very thin core conductor.

### D. All about Negative Capacitance

If the previous section has been read and understood, the mysteriously named "negative capacitance" will not cause perplexity. We have seen that the distortion or error in the recorded signal is due to a transient flow of current through $R_{\mu E}$ and $C_{tot}$, and that nothing can be done to stop the current from flowing through $C_{tot}$. But there is nothing to prevent us from supplying the error current from another source, so that it does not have to be drawn from $R_{\mu E}$. If the correcting current source were clever enough to supply precisely the right current $C_{tot}dE'/dt$, then $E$ and $E'$ would be identical. This is exactly what negative capacitance attempts to do.

In Fig. 13(a) the driven shield has been removed and two extra components added: an amplifier of adjustable gain $A'$ and a feedback capacitor $C_f$. The potential across $C_f$ is $A'E' - E' = E'(A' - 1)$ and so, from the definition of capacitance (Exercise 9 of Appendix I), the current through $C_f$ is $C_f(dE'/dt)(A' - 1)$. Since the desired correcting current is $C_{tot}dE'/dt$ we have only to adjust $A'$ until $C_f(A' - 1) = C_{tot}$. $A'$ is typically adjustable between 1 and 5 and $C_f$ has a fixed value of 2–5 pF. Substitution of these values shows that compensation for $C_{tot}$ up to 8 pF or 20 pF would be available.

We have managed to explain capacitance compensation without using the concept of negative capacitance. An alternative, but to my mind less convincing, explanation would be to say that the effective value of $C_f$ in Fig. 13(a) is $C_f(1 - A')$; since $A'$ is greater than 1 the effective value is negative. $C_{tot}$ is in parallel with the negative capacitance, and from the rule for combining capacitances in parallel the total effective capacitance is the algebraic sum of the two. Suitable adjustment of $A'$ brings the sum to zero.

How should the negative capacitance control be adjusted? In any low speed recording application it should be turned fully off, and left there. In high speed applications there are two main methods for adjustment. The first method can be used with any negative capacitance preamplifier, but unfortunately fails to give a single unambiguous correct setting. The principle is shown in Fig. 13(a). A rectangular-pulse generator or stimulator supplies a pulse of 50–100 mV amplitude to the bath via the indifferent electrode and the output signal is viewed at fairly high sweep speed on the oscilloscope. Correct adjustment of $A'$

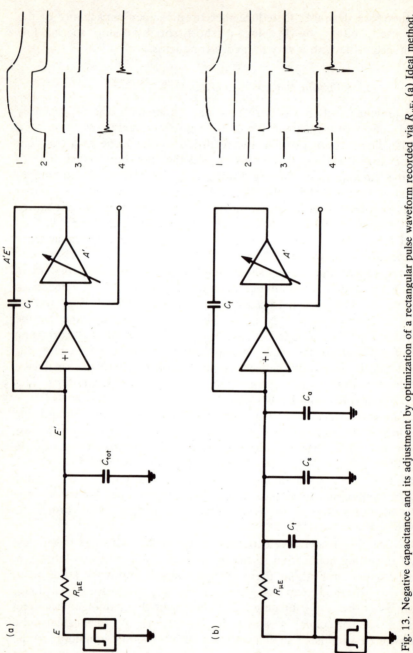

Fig. 13. Negative capacitance and its adjustment by optimization of a rectangular pulse waveform recorded via $R_{\mu E}$. (a) Ideal method. Waveforms 1 and 2 are undercompensated, 3 is fully compensated, and 4 shows the signs of overcompensation: overshoot and "ringing". Further increase in compensation will make the amplifier oscillate. (b) The flaw in the method. See text for discussion. Waveform 2 looks right but is actually undercompensated. Waveform 3 looks overcompensated but is actually correct.

gives the waveform marked 3. This method should be practised with a suitable resistor (10 MΩ or higher) in place of $R_{\mu E}$ and a 10 pF capacitor for $C_{tot}$. The flaw in the method is that the three components of $C_{tot}$ in the equivalent input circuit (Fig. 13(b)) are not connected in the same way, as seen by the stimulator. The transmural capacitance $C_t$ of the microelectrode is a capacitance to the bath (see Chapter 3, Section V), and the other two components are capacitances to earth. An approach to the correct adjustment is possible only if $C_t$ is made very small by keeping the depth of penetration of the microelectrode into the bath very small. Otherwise one has to guess at the correct setting (number 3 in Fig. 13(b)).

A better method is to inject a current pulse of about 1 nA into the input circuit and adjust $A'$ to give optimal appearance of the resulting voltage trace (the waveforms of Fig. 13(a)). If the preamplifier is already equipped with current injection facilities, no extra components are required. If not, the capacitor and voltage ramp method of Chapter 5, Section II is probably the simplest. It appears to be best to compensate so as to give the fastest rise time *without overshoot*; the dire effects of over-compensation are stressed by Unwin and Moreton (1974).

Theoretical considerations in the application of negative capacitance have been treated rather fully by Guld (1962), Schoenfeld (1962) and Moore and Gebhart (1962). A bold oversimplification of some practical utility is that the fully-compensated rise time is approximately twice the geometric mean of the amplifier's rise time and the (uncompensated) input circuit's rise time. Thus an amplifier of rise time 1 $\mu$s could, with the aid of negative capacitance, reduce an input circuit rise time of 200 $\mu$s to about 30 $\mu$s. An amplifier ten times as fast could reduce it to 10 $\mu$s. This explains why designers seek to make the input stage of a preamplifier as fast as possible (M. V. Thomas, 1977).

### III. Joining Wires to Cells

#### A. Connecting to the Microelectrode

The microelectrode must be held mechanically under control of the micromanipulator, and a metallic conductor must make contact with the filling solution inside the stem. The two requirements can be satisfied separately as in Fig. 14(a), but the more elegant arrangements of Fig. 14(b) and (c) save time when changing microelectrodes. For recording from moving tissues like heart muscle, the microelectrode can simply be

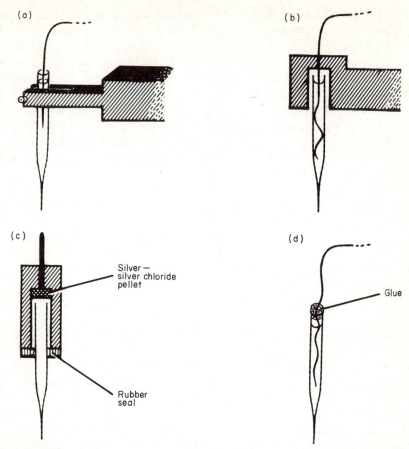

Fig. 14. Connecting to the microelectrode. (a) Separate mechanical and electrical connections. (b) Microelectrode retained in perspex holder by silver wire. (c) Microelectrode retained by rubber ring. The silver/silver chloride pellet connects to a metal pin that plugs straight into the preamplifier. Electrical continuity between microelectrode and pellet is maintained by an excess of filling solution previously injected into the holder. (d) Microelectrode dangled on flexible wire.

dangled on the end of a very thin wire (Fig. 14(d); see Woodbury and Brady, 1956) or plastic tube (Sato, 1977).

Electrical contact is usually made with chlorided silver wire or a sintered pellet of silver and silver chloride. The intention is to obtain a stable half-cell potential that will not drift during an experiment or be altered by the passage of small currents. These conditions can only be met by a *reversible* electrode. In a silver/silver chloride electrode the reversible reaction is $Ag + Cl^- \leftrightarrow AgCl + e^-$, the function of the silver

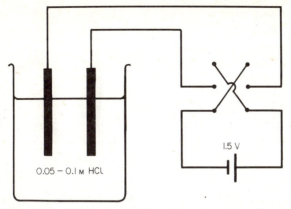

0.05 – 0.1 M HCl

1.5 V

Fig. 15. Chloriding silver wires or strips of silver. Clean them first with emery paper, then wash off any grease with alcohol. Mount as shown for about 20 min, operating the reversing switch every few minutes to alter the direction of current flow.

chloride being to provide a stock of chloride ions in solid form, ready for use should the reaction tend to proceed to the left. There are numerous recipes for chloriding silver (Janz and Ives, 1968). Thomas (1978) dips the wires in molten AgCl to produce a sturdy coating. A less troublesome method is shown in Fig. 15, but the grey or black Ag/AgCl coating formed is mechanically fragile and tends to flake off with use. The life of the coating is extended if the micropipette stems are fire-polished to remove the sharp edge (hold the stem in a gas flame for a few seconds, *before* filling the pipette). It should be mentioned that silver in contact with 3 M KCl becomes partly chlorided spontaneously, so that lazy people, or those with small interest in the recording of stable potentials, can save a little trouble by neglecting to chloride their wires.

### B. Connecting to the Extracellular Medium

The six methods shown in Fig. 16 do not exhaust the possible combinations, but will serve to illustrate the chief principles. The actual interface between metal and electrolyte solution is nearly always an Ag/AgCl electrode which may be a chlorided silver wire or piece of chlorided silver foil, or a sintered Ag/AgCl pellet. The latter offers a large surface area in compact form, but requires cleaning from time to time with fine sand-paper, especially if it comes in contact with proteinaceous solutions.

Method (a) in Fig. 16 is direct, obvious and widely used. It is

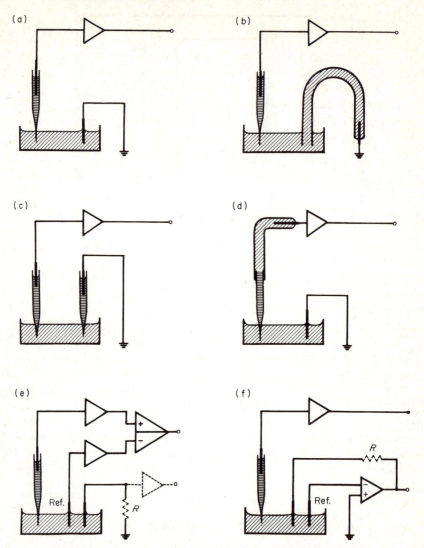

Fig. 16. Connecting to the extracellular medium. See text for discussion.

electrochemically asymmetrical in the sense that the two electrodes involved see solutions whose chloride activities differ. Since Ag/AgCl electrodes are reversible to chloride ions the preamplifier will record a steady offset potential. If the chloride concentration in physiological saline ("Ringer") is 150 mM and that of the microelectrode's filling solution 3 M, the offset potential will be $-26 \ln [3/0.15]$ mV $= -78$ mV with respect to earth (Geddes, 1972). A constant offset potential is easily compensated (Section IV of this chapter), but if the temperature of the bath changes, or if the chloride activity of the Ringer solution is altered during an experiment, the recorded resting potentials of cells will be in error.

Method (b) removes the indifferent electrode to a safe distance from variations in the temperature or composition of the Ringer solution, by placing it at the end of a Ringer – agar bridge 5–10 cm long. Ringer – agar is made by stirring 4% by weight of agar – agar into Ringer solution and heating gently until it dissolves. It can then be injected into a plastic tube containing the indifferent electrode, and allowed to cool. The purpose of the agar is to prevent bulk flow within the tube. If the ionic composition of the Ringer in the bath is changed, allowance must be made for the liquid junction potential between the bridge and the bath.

Electrochemical symmetry is provided by the arrangements of Fig. 16(c) and (d), which should show only a very small offset potential, apart from a possible tip potential of the microelectrode (Chapter 3, Section VI). In (c) a broken-tipped pipette filled with 3 M KCl forms a salt bridge between the reference electrode and Ringer solution, whereas in (d) a Ringer – agar bridge is interposed between microelectrode and preamplifier. Method (d) is not recommended for high speed recording because of the inevitably large stray capacitance of the Ringer – agar bridge. Screening against alternating electric fields is also a problem with this method.

The indifferent electrode in methods (a)–(d) really has two functions: to provide a stable reference potential, and to offer a return path for current flow entering the bath elsewhere. The current may come from a microelectrode used for current injection (Chapter 5) or ionophoresis (Chapter 6), or it may result from electric field interference. In either case it is important that the potential measured by the preamplifier is not altered by current flow through the indifferent electrode. The passage of any current whatsoever changes the potential of an electrode, but if the current is small, the surface area of the electrode large, and the electrolyte concentration high, the resulting potential change may be insignificant. The use of methods (a)–(d) presupposes that these conditions are met. If there is any doubt about this, methods (e) or (f) must be used. In both (e) and (f) the two functions of an indifferent electrode are assigned to separate electrodes. For simplicity the two electrodes are shown as simple Ag/AgCl electrodes, but either or both could have Ringer – agar bridges or a broken-tipped pipette interposed. The electrode labelled "Ref." connects to a high impedance input stage and therefore has virtually no current flow through it. The other electrode is the current return; during the passage of current an error potential develops across it but is prevented by the electronics from appearing at the output of the preamplifier. In (e) a differential amplifier subtracts the error, whereas in (f) an operational amplifier clamps the bath to the potential sensed by the reference electrode. Method (f) is recommended for providing a stable bath potential when several recording microelectrodes are in use. The resistors $R$ shown in dotted lines allow the current flow to be monitored, as discussed further in Chapter 5, Section III. If this feature is not needed, the resistors should be omitted and replaced by a direct wired connection.

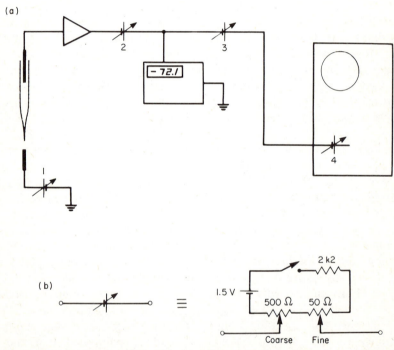

Fig. 17. Offset voltage controls. (a) An offset control at either 1 or 2 is used to compensate for tip potentials and other electrochemical asymmetries. The adjustment is made before impalement so that the digital voltmeter reads 0 mV; subsequently the voltmeter registers the resting potential. An offset control at either 3 or 4 compensates for the resting potential and allows the oscilloscope to be used at high sensitivity, without the trace disappearing off-screen. Control 2 is present in most preamplifiers; control 4 may exist within the oscilloscope's vertical amplifier. (b) A possible circuit for offset control, having COARSE and FINE adjustments.

## IV. Offset Voltage Control

Distinguish carefully between an offset control and the "position" control of the oscilloscope's vertical amplifier. The offset control alters the potential seen by the input to the amplifier, whereas the position control moves the trace up and down on the screen. The difference between them becomes apparent when the sensitivity of the oscilloscope amplifier is changed. Offset controls are used to compensate for unwanted steady potentials. If a digital voltmeter is available for monitoring resting potentials, it will be convenient to have two separate offset controls (Fig. 17).

## V. Earthing and Interference

The prime function of earthing is *safety*: protection against fatal electrocution. Its other functions — to provide a reference potential for the electronics and to screen against interference — must never be allowed to take precedence. Earthing can only protect if it is backed up by proper fuses in the live conductor of the main supply (Fig. 18).

Interference includes a wide variety of unwanted signals (Donaldson, 1958; Wolbarsht, 1964; Ott, 1976). Attention here will be directed towards prevention of interference from the mains power supply and especially to some aspects that the author finds poorly explained elsewhere.

Design of the earthing system begins with the choice of some physical

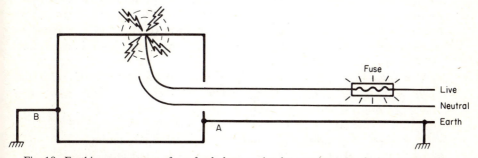

Fig. 18. Earthing protects you from fatal electrocution by carrying away fault currents and blowing the fuse in the live conductor. The metal case of every mains-operated piece of equipment must be permanently joined to earth with heavy-gauge wire. Connection A is safest. In special circumstances A might be removed, but only if replaced by a secure connection B. *Ordinary hook-up wire temporarily plugged into an earth socket is not good enough.*

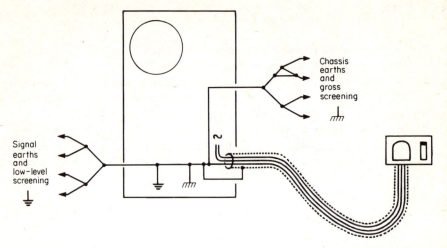

Fig. 19. The oscilloscope forms a useful base for the earthing system. Note the different symbols for chassis earth and circuit earth. There must be one and only one connection between them. Chassis earths may have multiple interconnections. The power cable to the oscilloscope is screened to contain its alternating electric field.

piece of metal and the *definition* of its potential as "earth" or 0 V. It is immaterial whether or not its potential really is zero with respect to an external reference or abstract ideal "earth"; the whole system could be at an (alternating) potential of, say, 100 mV *provided that every part of the apparatus regarded it as zero.* Some texts recommend the use of a water pipe or copper rod hammered into the ground. This approach usually leads to the removal of all the mains earths, and unless properly engineered may be unsafe. My recommendation (Fig. 19) is to leave the mains earth connected to the oscilloscope and to choose the oscilloscope's frame as 0 V.

A distinction of great importance is made in Fig. 19 between *chassis earth* and *signal earth*. Chassis earth includes the metal cases of mains-powered equipment; its first job is safety, but it also provides crude screening against alternating mains voltages. Signal earth gives a reference (0 V) potential to amplifiers and the like. For example, the indifferent electrode in Fig. 16 is often connected directly to signal earth, forming a reference for the preamplifier. To avoid *earth loops* (Section C below) there must be one and only one connection between chassis earth and signal earth. If the recommendation of Fig. 19 is followed, this single connection will be at the oscilloscope.

### A. *Alternating Electric Fields*

Capacitive coupling between the power lines and the microelectrode input circuit is responsible for most of the interference problems in intracellular recording, because of the very high impedance of the input circuit. Figure 20 shows how the interfering signal is transmitted from the source "aerial" through the dielectric medium of the air with the high impedance part of the input circuit acting as a "receiving aerial". There are four remedies: make the "aerials" as small as possible, remove the alternating mains voltage from the source, increase the physical separation between the source and the input circuit, and interpose an earthed conductive shield. The general rule is obvious: any conductor which can be "seen" by the input circuit must have a small or zero alternating voltage on it *or* be physically distant (at least several metres) *or* be physically small.

From the standpoint of interference reduction it does not matter much

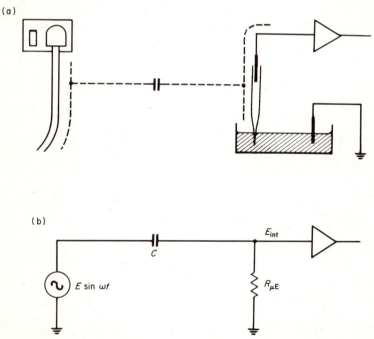

Fig. 20. Interference from alternating electric fields. (a) Capacitive coupling between source aerial and "receiver" aerial. (b) Equivalent circuit. The interfering source is at potential $E \sin \omega t$. The interference $E_{int}$ at the receiver is given approximately by $E \omega R_{\mu E} C \cos \omega t$, the approximation being valid because $C$ is generally only a fraction of a picofarad. Note that $E_{int}$ is proportional to $R_{\mu E}$; this is of great diagnostic help.

where the earthed shield is placed. Often it is most convenient to shield the source; a good start to clean up the hum level can be made by enclosing every mains power cable in an earthed metal braided screen. The fully enclosed Faraday cage of yesteryear is unnecessary and inconvenient. However, a metal enclosure built around the experimental set-up and having one side completely open for access is very useful, provided that it is not regarded as an ultimate weapon against alternating electric fields. The idea of such an arrangement is that it eliminates five of the six possible directions for capacitive pick-up to reach the input circuit, and so when interference is experienced it will be known to come either through the sixth side or via some non-earthed conductor which enters the enclosure. I have a large (2 m × 2.5 m × 3 m) enclosure screened on five sides with expanded aluminium mesh.

The question is often raised whether mesh or chicken wire is as effective for screening as a continuous metal sheet. Suppose the holes in the mesh to have a width $x$, and suppose the distances between source and screen and between screen and "receiver" to be $y$ and $z$. Then the mesh behaves as a continuous sheet if $y \gg x$ and $z \gg x$.

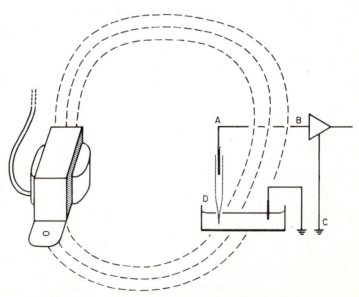

Fig. 21. Interference from alternating magnetic fields. From the laws of electromagnetic induction, the interfering e.m.f. is proportional to the magnetic flux intensity and to the area ABCD intersected by the alternating flux.

### B. Alternating Magnetic Fields

Transformers (especially cheap ones) and electric motors produce large alternating magnetic fields in the space near by. If flux lines cut part of the signal circuit they will induce an alternating e.m.f. in it. The interference is proportional to the area of the signal circuit (ABCD in Fig. 21), and normally will be very small unless the earth return lead from the preparation has been laid out in a circuitous route.

Attempts at magnetic shielding are unlikely to be rewarding, and the best remedies are to remove the offending source and to reduce the area of the signal circuit. Microscope lamps and their associated wiring and transformer commonly cause both magnetic and electric field interference. A d.c. power supply for the lamp filament overcomes these problems; it should be well filtered but need not be regulated.

### C. Earth Loops

Beginners are often unpleasantly surprised to discover that not all wires and terminals labelled "earth" are at the same potential. But wires and connectors have a finite resistance, and if an alternating current flows through them an alternating potential must be developed. The commonest cause of this trouble is the earth loop (Fig. 22). In Section B above it was noted that alternating magnetic fields rarely interfere with

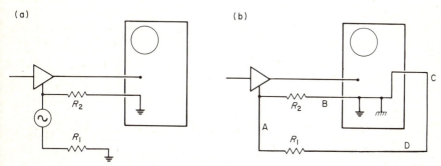

Fig. 22. How earth loops cause interference. (a) Earth wires and connectors have a finite resistance (a fraction of an ohm) which can couple an interfering e.m.f. into the signal circuit. (b) The commonest interfering e.m.f. is that due to alternating magnetic fields intersecting earth loop ABCD. Note that the loop has been formed by a connection at A between chassis earth and circuit earth. Connection A should be removed to break the loop. These diagrams explain one of the most puzzling features of earth-loop interference: connection of extra earth wires sometimes increases the hum level and sometimes reduces it. A wire in parallel with $R_1$ will increase the loop current through $R_2$ and worsen the interference; a wire in parallel with $R_2$ also increases loop current but the voltage across $R_2$ is reduced and the hum made smaller.

the input circuit proper, but the area enclosed by an earth loop may be hundreds of times that of the input circuit, with a corresponding increase in the interfering electromotive force.

Dealing with earth loops can be a tedious business. In the trivial example of Fig. 22(b) the problem is solved by breaking of the loop at point A. More difficult problems are posed by the loops created when several amplifiers are powered from a common d.c. supply (case history 2 below). The most difficult, but least talked about, problems result from the earthing design of certain commercial items of equipment. Figure 23 shows that any attempt to pass a signal between two mains-powered units that have a combined chassis-earth/signal-earth produces a loop. Some manufacturers are kind to the user, and separate their earths, e.g. at point A of Fig. 23, so that the distinction between chassis earth and signal earth can be maintained. Those that are unkind can justify their design with the argument that it is safer to have signal earth permanently joined to mains earth. Undoubtedly the argument is sound, but most users, faced with the loop in Fig. 23, elect to break it at point E so that there is now no safety earthing at all! This is the origin of the thoroughly bad advice sometimes given to disconnect earth wires of sensitive equipment at the mains power plug.

The right way to fix this problem is to mount the offending apparatus on the same rack as the oscilloscope and to supply it from the same power socket or distribution board. Less satisfactorily, a heavy

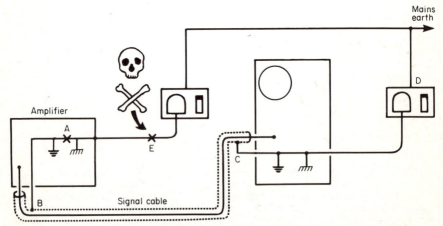

Fig. 23. Earth loop due to inconvenient earthing design of equipment. Both the oscilloscope and the amplifier at left have their circuit earth and chassis earth combined. Connection of a signal cable BC forms a large loop ABCDE. Separation of circuit earth from chassis earth at point A, as done by some manufacturers, interrupts the loop. Disconnection at E also breaks the loop but is unsafe.

permanent connection could be made between the oscilloscope frame and point A of Fig. 23, following the path in space of the signal cable BC; disconnection at E will now break the loop without undue risk of electrocution.

### D. What to Do if It Hums

Diagnosis should precede treatment. Consult Figs 24 and 25. Once the cause of interference is known, the cure is usually obvious.

### E. Three Case Histories

*1. Hum current through microelectrode: manufacturers fault.* A brand new electrometer amplifier was bought from a well-known American manufacturer and in preliminary trials appeared to work satisfactorily. Some days later it was used in an experiment which required microelectrodes of resistance 100–150 MΩ. A hum voltage of about 0.5 mV peak-to-peak became apparent. Cursory investigation according to Section D above suggested that alternating electric fields were responsible, since the hum level dropped to only 5 μV when the microelectrode was short circuited. As expected, a 100 MΩ resistor gave 0.5 mV hum. Some hours were wasted in trying to locate the source of the interfering field. Then it was found that local screening of the input circuit failed to eliminate the hum. Attention was therefore turned to the preamplifier itself, since the evidence implied that a hum *current* of about 5 pA peak-to-peak was flowing in the input circuit; the resulting hum voltage was naturally proportional to the resistance through which the current was made to flow.

After a number of false trails had been followed, the culprit was discovered to be the power transformer in the main amplifier chassis. This transformer had an unusually high leakage inductance which gave rise to a strong alternating magnetic field, some of which intersected the adjacent circuitry. The local agent for the amplifier was reluctant to believe in this defect, but was convinced by direct demonstration. A complete cure was effected by removing the transformer from the chassis and mounting it on an outrigger platform 20 cm behind the amplifier. Subsequent communication with the manufacturer showed that the problem occurred only in export models intended for 230 V mains operation, the transformer in the 110 V versions being of better design.

Two lessons may be drawn from this case history. The first is that even expensive commercial equipment may be defective. Sometimes the

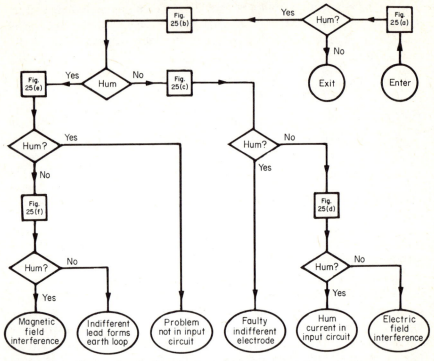

Fig. 24. Flow diagram for the investigation of interference at the mains frequency. The name "hum" derives from the characteristic sound when the interfering signal is played through an amplifier and loudspeaker.

defect is fairly obvious; for example a preamplifier of British make was received from the manufacturers with gross electrical misalignment and showed a leakage current of 0.5 nA. Sometimes, as the case history shows, the defect may be far from obvious. The second lesson is that agents and manufacturers are reluctant, and in general rightly so, to believe complaints about hum level. They receive many such complaints and nearly all of them turn out to be the client's fault (earth loops and the like), as exemplified below.

*2. Hum current through microelectrode: user's fault.* Preamplifiers with current injection facility (Chapter 5) provide an input terminal which accepts a voltage signal and converts it into a proportional current flow through the microelectrode. An ordinary stimulator was used to supply the voltage signal; for various reasons it was thought desirable to have the stimulator on one equipment rack and the preamplifier on another. The output of the stimulator was earth-referenced. When the

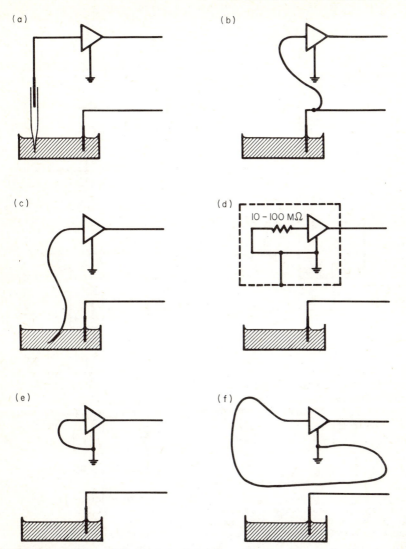

Fig. 25. Arrangements of the input circuit for diagnosis of interference (hum) at the mains frequency. See instructions in Fig. 24.

connections were made (Fig. 26(a)) a hum current, diagnosed as in Section D above, of about 20 pA peak-to-peak was observed in the microelectrode circuit. The hum was made worse by removing the earth connection from the screened stimulus signal cable at A or B, and somewhat better by connecting another earth lead in parallel with the cable's screen. Evidently the alternating potential at earth A on one

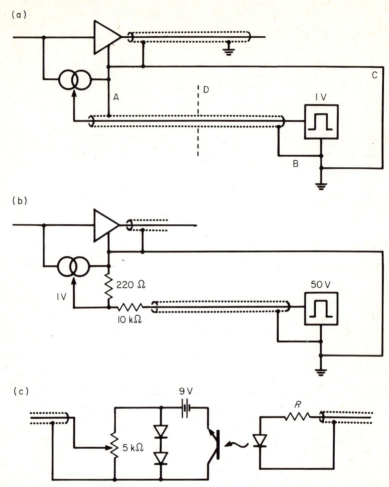

Fig. 26. Earth loop in stimulus input circuit, causing hum current through microelectrode. (a) Original connection, showing earth loop ABC. Lead C was permanently wired and could not be removed. Breaking the loop at A or B did not cure the problem since the stimulus signal still contained hum voltage. (b) The stimulus signal is increased 50 times to get it well above hum level, and then divided by ~50 at the preamplifier. (c) The loop is broken with an optical isolator circuit, interposed at point D in diagram (a). Resistor $R$ should be selected to give an output current $I$ from the phototransistor of 1–2 mA. The silicon diodes limit the signal to ~1.2 V.

equipment rack was a millivolt or so different from that at B on another rack. Possible solutions: (1) place stimulator and preamplifier on the same rack; (2) increase the stimulus signal above hum level (Fig. 26(b)); (3) use a stimulator with isolated output; (4) make a simple isolator circuit (Fig. 26(c)). All four solutions worked, and the last-mentioned was the one actually used.

The obvious lesson here is that not all terminals labelled "earth" are at the same potential. The problem could have been avoided altogether if the isolator had been designed into the preamplifier (Appendix IV).

*3. A tangle of earth loops.* A current monitor (Chapter 5, Section III) was built and installed so as to measure current flow in the return lead from the indifferent electrode during current injection or ionophoresis. Initially the power to the monitor was derived from a ±15 V supply which also powered a number of other (earthed) circuits. As expected, the resulting earth loops led to an intolerable hum voltage on the indifferent electrode and bath. Next, the power to the monitor was taken from the power supply of the preamplifier itself, the reasoning being that as far as the signal was concerned, the preamplifier's earth was 0 V by definition. The arrangement, in which three loops may be distinguished, is shown in Fig. 27. A large hum voltage was still present on the bath return lead. Disconnecting the screen at "E" lessened the hum slightly, but now several millivolts hum appeared on the current monitor's output. "E" was reconnected, and the power supply leads and cables BC and ED rearranged in space so as to minimize the loop areas. A complete cure was obtained.

The case history illustrates the difficulty of sorting out earth loops, and the mixed advantages which mains-operated power supplies confer. In extreme desperation it may be necessary to give each circuit its own battery supply.

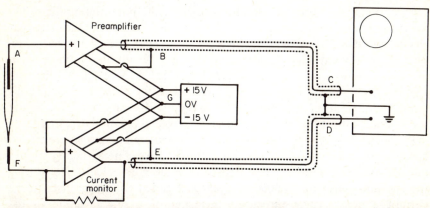

Fig. 27. A tangle of loops. A preamplifier and current monitor are connected to a common ±15 V power supply. Three loops are present: ABCDEF, BCDEG and AGF. Because the circuits are closely related and have a common power supply it is best *not* to break the loops but to arrange the leads in space so as to minimize the enclosed areas.

## F. Stimulus Artefacts and Impulsive Interference

Capacitive coupling to the microelectrode via the dielectric medium of the air allows it to pick up unwanted signals from stimulating leads or from adjacent microelectrodes to which voltage pulses are being applied. The principles outlined earlier for coping with interference from alternating electric fields are fully applicable to the elimination of impulsive interference of this kind.

More serious problems arise from other mechanisms of coupling between stimulus electrodes and the recording microelectrode. The current required to excite nerves with extracellular electrodes is measured in microamperes or even milliamperes, and it is of considerable practical importance to ensure that no more than a tiny fraction of it flows through the indifferent electrode. The latter, if well designed, may accept currents up to a few hundred nanoamperes without significant change of potential, but large stimulus currents will disturb it and introduce an unpleasant stimulus artefact which may take several seconds to die away. This state of affairs is in practice invariably avoided by the use of a stimulator with isolated ("floating") output. Neither output terminal has a direct connection to earth, and so current emitted from one terminal returns to the other after traversing the stimulating electrodes and the tissue. In high gain recording the stimulator must have good isolation; in particular the capacitance to earth must be small. The signals recorded in intracellular work are relatively large, so that less stringent demands are placed on the stimulus isolation.

The remaining source of stimulus artefact is the potential gradient due to current flow in the extracellular medium between the stimulus electrodes. If the recording microelectrode is near one of the latter, an artefact of several volts is almost certain to be observed. It is often possible to reduce this to an acceptable level merely by positioning the stimulus electrodes a long way from the recording electrode. When this cannot be done, one may have to be content with a large artefact during the stimulus pulse; this need not obscure subsequent cellular responses if matters can be arranged so that the recorded signal returns swiftly to baseline after the pulse. A delayed return to baseline is nearly always due to continued current flow between the stimulus electrodes: during the stimulus pulse these become polarized and then slowly discharge a current in the reverse direction. Inclusion of a silicon diode in the stimulus circuit helps to prevent reverse current flow. Use of a constant current stimulator in place of the more usual constant voltage type is equally effective. (N.B.: "constant" current does not mean that the stimulus cannot be varied; it means that the current is substantially independent of the nature of the electrodes, polarizing voltages and so on.) Another cause of delayed return to baseline is injudicious use of a low-pass filter (see Fig. 30(b)).

A completely different kind of impulsive interference is that arising when electrical appliances are switched on or off. Thermostats cause most of the problems, and some detective work may be called for to discover which of the refrigerators, water-bath heaters, laboratory ovens or chromatographic chambers in the building is responsible for one's present trouble. The suppression of switching artefacts is difficult, and if the offending appliance cannot simply be turned off, expert advice should be sought.

## VI. Noise

Noise is sometimes defined as "unwanted signal". I do not much like this definition since it obscures the fundamental and extremely important distinction between interference, as discussed in Section V of this chapter, and random noise. "Noise" in the present sense refers to unpredictable spontaneous fluctuations that result from the physics of the substances making up the electrical circuit. Most kinds of electrical noise are a consequence of the quantal nature of electric charge. The importance of noise to the electrophysiologist is twofold: (1) noise in the input circuit sets a limit to the smallness of signals that can be reliably measured, and (2) analysis of noise in biological signals can reveal something of the molecular events in cellular electrogenesis. The following discussion is restricted to case (1). A more advanced treatment of noise in electronics is given by Motchenbacher and Fitchen (1973).

The noise performance of a system can be expressed in many different ways, few of which are immediately revealing or useful to the electrophysiologist (e.g. signal-to-noise ratio, noise figure). The amplitude of the noise voltages is of more direct relevance. But because the voltage varies unpredictably we have to introduce some sort of time average value. The arithmetic mean noise voltage is, however, zero since fluctuations of either sign are equally probable. If the measured noise voltages are squared, and the mean square over some suitable long interval of time is computed, fluctuations of both signs contribute positively. The square root of this mean (root mean square or r.m.s., $\hat{E}$) is the usual way of expressing the magnitude of noise voltages. Separate (uncorrelated) noise voltages sum according to the relation

$$\hat{E}_{\text{tot}} = (\hat{E}_1^2 + \hat{E}_2^2 + \ldots)^{1/2}. \tag{10}$$

Thus $\hat{E}_{tot}$ is the root-sum-square of the components.

The r.m.s. value $\hat{E}$ does not constitute a complete description of noise; we need in addition some measure of how rapid or slow the fluctuations are. This is provided by the power density spectrum, or equivalently by the autocorrelation function. The well-known "white" noise, of which Johnson noise is a prominent example (see below) has a flat power spectrum; total noise power is directly proportional to noise bandwidth. Since noise power is expressable as $\hat{E}^2/R$, where $R$ is the resistance of the circuit, a moment's thought shows that r.m.s. noise $\hat{E}$ is proportional to the square root of bandwidth. That is, if a system with a bandwidth of 0–10 kHz exhibits r.m.s. noise of 200 $\mu$V, extension of the bandwidth to 40 kHz will double the noise to 400 $\mu$V. Conversely, reduction of the bandwidth to 2.5 kHz will halve the noise to 100 $\mu$V. This is the reason for using low-pass filters (see Section VIII of this chapter).

Some kinds of noise, generically called flicker or $1/f$ noise, have predominantly low frequency fluctuations, and low-pass filters are less effective in reducing the r.m.s. value. Unfortunately the noise contributed by active devices like junction transistors, field-effect transistors and vacuum valves is partly of this kind.

When viewed on an oscilloscope at low sweep speed, noise appears as a thickening of the baseline. An unwelcome fact is that the thickness is five to eight times the r.m.s. value. If the r.m.s. noise is 200 $\mu$V the baseline will appear 1–1.5 mV wide (depending on the spot intensity), with occasional peaks above and below. Measurement of miniature end-plate potentials would be very inaccurate with this noise level.

There are four chief sources of noise in the input circuit for microelectrode recording: (1) Johnson noise of the microelectrode's resistance, (2) the preamplifier's voltage noise, (3) the preamplifier's current noise, (4) "excess" noise in the microelectrode.

(1) Johnson noise is due to thermal motion of electrons, and sets a lower limit to the total noise. The r.m.s. value is $(4kTR\Delta f)^{1/2}$, where $k$ is the Boltzmann constant, $T$ is the absolute temperature, $R$ is the resistance of the source, i.e. the microelectrode, and $\Delta f$ is the noise bandwidth. If $\Delta f$ is 10 kHz this formula gives 40 $\mu$V for a resistance of 10 M$\Omega$ and 130 $\mu$V for a resistance of 100 M$\Omega$, the temperature being 20 °C = 293 K in both cases. Johnson noise is automatically band-limited by the low-pass filter consisting of $R_{\mu E}$ and the total input capacitance.

One commercial preamplifier claims a *total* noise less than the room-temperature Johnson value appropriate to the resistance used in the test, the value quoted suggesting that the measurement must have been made at $-40\,°C$! Although we must reject this claim as extravagant, it remains true that a good preamplifier need not contribute significantly to the overall noise. Over a certain range of source resistances (perhaps 10–50 MΩ) the observed noise may be substantially that of the source. However, at very low or very high source resistances the preamplifier's voltage noise and current noise respectively become the dominant noise generators.

(2) Voltage noise inherent in the preamplifier is the noise measured at the output when the input is earthed. Typical values are 5–20 $\mu V$ r.m.s. at 10 kHz bandwidth and form an insignificant contribution to the total noise. In an exceptionally good design the voltage noise might be only 1–2 $\mu V$. The preamplifier's voltage noise, as such, is hardly ever of importance since in practice it is dominated by the microelectrode's Johnson noise.

(3) All preamplifiers produce some current flow in the input lead. The steady component (d.c. bias) is considered in Section II.A of this chapter. Superimposed on this are random fluctuations, having some r.m.s. value $\hat{I}$. The noise current flows through the parallel combination of microelectrode resistance and stray capacitance, producing a noise voltage $\hat{E} = \hat{I}R_{\mu E}$, which is band-limited as in (1) above. Current noise is noteworthy for its linear dependence on $R_{\mu E}$, contrasting with the square root dependency shown by Johnson noise. As $R_{\mu E}$ is made larger, eventually current noise will come to dominate the total noise. The manufacturers of microelectrode preamplifiers never quote values of noise current; 0.2–0.5 pA r.m.s. is perhaps a reasonable guess for the better designs. The worst designs may have several picoamperes. If $\hat{I}$ is 1 pA, current noise will equal Johnson noise for a resistance of 160 MΩ. The lesson from this is that the preamplifier needs to have low noise current (see below) if you need low noise performance with very high resistance microelectrodes such as ion-sensitive types (Thomas, 1978).

(4) Microelectrodes show a noise component additional to their Johnson noise. This "excess noise" depends strongly on the voltage applied to the microelectrode, and therefore increases markedly during current injection. Some excess noise is present even in the absence of applied voltage (DeFelice and Firth, 1971).

We noted above that preamplifier voltage noise, as such, is usually too small to be important. Unfortunately the more complicated preamplifiers have a circuit configuration which converts voltage noise into noise current. In the simplified schematic of Fig. 28 is shown a unity-gain amplifier with voltage noise $\hat{E}$ and feedback impedance $Z_f$. The circuit could represent a current-pump type of preamplifier with no command signal applied; in this case $Z_f$ is the resistance $R_s$ (Chapter 5,

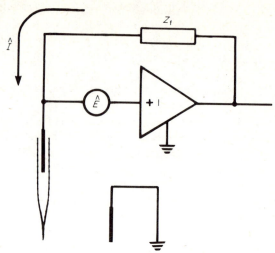

Fig. 28. Feedback via impedance $Z_f$ across a unity-gain amplifier converts the amplifier's voltage noise $\hat{E}$ into noise current $\hat{E}/Z_f$.

Section II). Or it could represent a preamplifier with negative capacitance abilities, adjusted so that the negative capacitance is off. In this case $Z_f$ is the impedance of a capacitance $C_f$ of a few picofarads (Section II.D in this chapter). The circuit of Fig. 28 is more or less identical with that of the current pump (Fig. 31(f)) discussed in Chapter 5, Section II and therefore generates a noise current $\hat{I} = \hat{E}/Z_f$. The designer of preamplifiers is (or ought to be) well aware of this source of noise current, and aims for low $E$ and high $Z_f$, that is high $R_s$ or low $C_f$. From the user's point of view there is not much which can be done to improve a particular design, but for the ultimate in low current noise a simple preamplifier without feedback components is to be preferred. We have discussed $Z_f$ as if it were internal, and thus the designer's responsibility. But it might equally well be the capacitance to core of a driven screen (Section II.C in this chapter), which is certainly the user's responsibility. There are good reasons for using a driven screen, but low current noise is not one of them. The capacitance should be kept small by using special low capacitance cable of the shortest possible length.

The noise performance of a preamplifier with negative capacitance is difficult to analyze theoretically, and also difficult to measure in a useful way, since the observed noise depends critically on the fine adjustment of negative capacitance. A greatly simplified analysis shows that with full compensation the noise contribution is proportional to $(C_{tot} + C_f)$, where $C_{tot}$ is the total stray capacitance and $C_f$ the feedback capacitance. This shows that it pays to keep $C_{tot}$ small by good layout. It also pays the designer to keep $C_f$ small. In M. V. Thomas' design (1977) $C_f$ is not present at all, negative capacitance being applied through the input capacitance $C_a$ of the amplifier itself.

The effect of negative capacitance on noise can be seen by connecting a 10 MΩ resistor in place of the microelectrode and observing the preamplifier's output at high sensitivity (say 0.1 mV per division) on an oscilloscope. As the negative capacitance control is advanced from its off position, the displayed noise increases dramatically. The increase is partly due to an improvement in the amplifier's ability to transmit the Johnson noise of the resistor, and partly due to additional current noise introduced via the feedback capacitance $C_f$.

## VII. Measuring Microelectrode Resistance

Remember that the resistance of a microelectrode is not a constant but varies with current (Chapter 3, Section VII). Great accuracy of measurement is unnecessary, an error of 20% being of little consequence.

The *constant current method* is illustrated in Fig. 29. It is a fast no-nonsense method, especially if the current source is combined with the preamplifier and operated by a push-button (Appendices III and IV) to give a steady current of 1 nA. With a digital voltmeter for readout, resistance measurements can be made in less than a second. With slight loss of convenience, the 1 nA current can be supplied as a brief pulse by

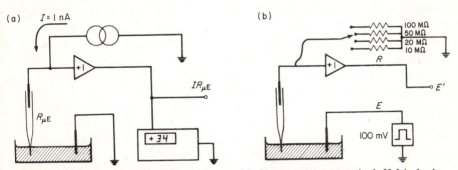

Fig. 29. Measuring microelectrode resistance. (a) Constant current method. If $I$ is 1 nA, the resistance in megohms is simply the number of millivolts shift when $I$ is turned on. (b) Voltage divider method. A temporary jumper lead is joined to the preamplifier input and to a known resistance $R$. With the symbols shown, the calculation to give microelectrode resistance is $R_{\mu E} = RE'/(E - E')$. The accuracy of the measurement is improved if $R$ and $R_{\mu E}$ are of the same order, hence the range of values of $R$, switched as indicated.

the current source of Fig. 31(c), rather than as d.c. For practical reasons the pulse can only be a few milliseconds in length and so an oscilloscope must be used to view the resulting voltage pulse. A variant of this idea is to apply a triangular waveform to the small capacitor of Fig. 31(c). Further details are given by Thomas (1978).

The *voltage divider method* of Fig. 29(b) requires you to work through some mental arithmetic to arrive at the resistance. If you use more than a dozen microelectrodes in a day, this method will become very tedious.

In low noise recording applications it is undesirable to have extra leads or components permanently attached to the input. The voltage divider method is usually brought into play via a temporary jumper lead as shown in Fig. 29(b). There is no reason why the constant current method should not be applied in the same way, the jumper lead being removed after use.

## VIII. Low-pass Filters

The inadvertent, and often unwanted, low-pass filter existing in the input circuit for intracellular recording has been discussed in Section II.C of this chapter, with the intention of showing how its harmful effects on rapidly changing signals could be mitigated. In low speed recording applications it is usually desirable to leave this low-pass filter alone, and even to introduce extra filtering.

The name "low pass" derives from the behaviour of the filter when it is fed with sinusoidal signals of different frequencies: signals of sufficiently low frequency pass to the output unchanged, whereas progressively higher frequencies are attenuated more and more (Fig. 10(b)). In Fig. 30(a) $R$ and $C$ form a voltage divider in which the impedance of $C$ decreases with increasing frequency. High frequency signals are thus shunted to earth and do not reach the output. The circuit of Fig. 30 is the most primitive type of low-pass filter but is suitable for routine work.

The filter has two main functions. One is to remove radio-frequency interference if this is present. The other is to reduce the noise level (Section VI in this chapter). A filter cannot distinguish between wanted and unwanted components of its input signal, except by their rate of change. If the filter's rise time $t_r$ is too long (or, equivalently, its cut-off frequency too low) the rapidly changing parts of the wanted signal will be distorted. A multiway switch allows choice of the appropriate filter characteristic to ensure that $t_r$ is at most a fifth of the signal's time to peak.

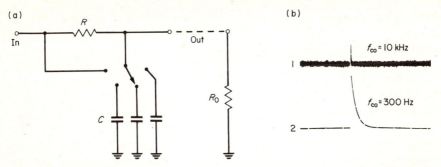

Fig. 30. Low-pass filter. (a) Simple circuit for connection between preamplifier and oscilloscope. A switch selects desired filter characteristic or OFF position. $R$ must be much less than the input resistance $R_0$ of the oscilloscope or else the filter will introduce a low frequency gain error. $R_0$ is typically 1 M$\Omega$ so that $R$ can be 10 k$\Omega$ with a gain error of about 1%. If $R = 10$ k$\Omega$ and $C = 1.5$, 5.6 and 15 nF, the filter rise times are 33, 120 and 330 $\mu$s, roughly corresponding to cut-off frequencies of 10, 3 and 1 kHz. (b) Effect of filtering on noise and spike voltage interference. Reduction of bandwidth from 10 kHz to 300 Hz virtually eliminates noise but makes the spike appear more prominent. Trace length: 20 ms.

Paradoxically, a filter may worsen the appearance of some forms of interference. Fig. 30(b) shows how interfering spikes, due perhaps to switching artefacts, are reduced in amplitude but smeared out in time.

Some preamplifiers are fitted with a "notch filter" centred on the frequency of the mains electricity supply (50 or 60 Hz). Notch filters are apt to produce bad distortion of waveforms and my advice is not to use them.

## IX. Calibration

The physical quantities of interest are voltage, time, current, resistance and capacitance. It is rare for any of these to need measurement with an absolute accuracy better than a few per cent. Considerable laxity in the matter of calibration is therefore permissible, especially in view of the excellent stability of modern electronic apparatus.

The least troublesome "calibration" method is simply to believe the values printed on the oscilloscope's time-base and vertical amplifier controls. If the instrument is returned periodically to the manufacturer for recalibration, this method may well prove adequate. I prefer to supplement it by regular checks on the oscilloscope's performance with the aid of a fairly accurately known voltage and a crystal-controlled oscillator. A Mallory mercury voltage-reference cell provides a sufficiently accurate potential of 1.35 V if the current drawn from it is not allowed to exceed 100 $\mu$A. Alternatively, a digital voltmeter offers a very high standard of potential measurement.

Some workers find it convenient to include on every oscilloscope trace a calibrating pulse of, say, 1 ms duration and 10 mV amplitude. This method is really a hangover from the days when vacuum tube amplifiers of uncertain gain were used, but it is still useful for keeping track of the amplification of a signal that undergoes many stages of processing. It is quite difficult to generate pulses of precisely known amplitude; the markings printed on the front panel of a stimulator are notoriously inaccurate.

Time calibration may be carried out with reference to the alternating mains frequency (50 or 60 Hz) whose long term accuracy is good, as exemplified by electric clocks, but which may undergo short term fluctuations of a few per cent. A better alternative is a quartz crystal-controlled oscillator with logic divider chain, giving output pulses at decade multiples of 1 ms.

Calibration of resistance is made by comparison with precision resistors. Values up to 1 M$\Omega$ are readily available with 1% tolerance. Calibration of current can usually be reduced to the measurement of voltage across a known resistance. Most commercial capacitors have tolerances of 10% or worse, but 1% types are available for the rare occasions when calibration of capacitance is required.


## X. Trouble-shooting and Fault-finding

Equipment failure is sufficiently common that the electrophysiologist needs to be continuously aware of the possibility. When things go wrong it will usually be because incorrect connections have been made or controls set to the wrong position, and only occasionally because of hardware failure. When a malfunction is first noticed the cause may not be evident at once. The important thing then is to devise some diagnostic tests that will locate the offending piece of equipment.

Successful diagnosis requires (1) a thorough knowledge of the expected normal function of each item, (2) some signal sources and display devices to replace those suspected of being faulty, and (3) a logical plan. The level of logical analysis needed is not very high. For example if A is supposed to supply a signal to B but appears not to do so, the fault may lie with A or with B or with the connection between them. It will generally be possible to devise tests to examine the three cases separately. Once the fault is tracked down in this manner a careful check for mistakes can be made. (Is the power turned on? Are the cables plugged into the right sockets?) In this way hardware failure can be diagnosed by exclusion of other possibilities: the defective equipment can then be examined further or sent for specialist attention.

Interconnecting cables should be regarded with deepest suspicion, especially those whose plugs have been attached hastily in the middle of an experiment. Failure of a cable is usually all-or-nothing: either the central conductor becomes open circuit or it becomes shorted to the screen. Either way, no signal gets through and diagnosis is easy. The pretentiously named "low noise" cable may fail in a different way. Inside the metal braided screen of this type of cable is a slightly conductive layer which is intended to reduce the magnitude of the "triboelectric" effect (noise produced by friction when the cable is bent). Unless some care is taken, the conductive layer all too easily comes in contact with the signal-carrying central conductor and forms a resistive shunt to earth. The behaviour of the resulting fault condition may range from the inexplicable to the incredible, until one remembers to measure with an ohm-meter the resistance between central conductor and screen!

Some miscellaneous hints are offered below, to provide further illustrations of the art of fault diagnosis.

(1) If a problem appears in the input circuit or preamplifier, begin your testing by replacing the microelectrode with a piece of wire, or (more realistically) by a 10 M$\Omega$ resistor.

(2) Batteries are good enough power supplies for many purposes, but nothing is more certain than that they will become exhausted sooner or later. When testing them with a voltmeter, leave them connected and make the measurement with the circuit they supply switched on (a failing battery may give the right voltage on open circuit but drop under load).

(3) If four pieces of apparatus fail simultaneously, it is probably because their common power supply has died.

(4) Fuses rarely blow spontaneously — they blow because they have "detected" a fault. Try to find it before replacing the fuse.

(5) Electronic components with moving parts, like switches and variable resistors, are less reliable than fixed components.

(6) Electronic technicians need to know what a faulty equipment item is supposed to do so that they can fix it. They also need a circuit diagram. It's wise not to lose the instruction manual.

# 5
# Current Injection

## I. Introduction

There are several reasons for wanting to pass a current through a microelectrode: measurement of resistance (Chapter 4, Section VII), ionophoresis (Chapter 6), and "stimulation" of cells are the chief ones. Moreover, in many applications one wants to record the membrane potential of the impaled cell or a neighbour during the injection of current to determine passive electrical properties, reversal potentials for neurotransmitter action, or electrical coupling between cells. First we consider the ways in which a current can be made to flow through a microelectrode, more or less under the experimenter's control.

## II. Current Sources

*Direct connection to a voltage source* (Fig. 31(a)) is the most brutal method of passing current through a microelectrode. For our purposes the method has nothing to recommend it. Even when no voltage pulse is applied, current can leak out of the impaled cell, through the microelectrode and stimulator to earth. And when a voltage is applied, the resulting current will depend on the (non-linear) resistance of the microelectrode.

The *series capacitor method* (Fig. 31(b)) improves matters a little. The capacitor ($0.01–0.1$ $\mu F$) ensures that the current averaged over a long time is zero, and thus prevents leakage of current out through the microelectrode. But by the same token it prevents extended currents from being applied to the cell; the actual current waveform is rather bizarre. Fatt and Katz (1951) used this method to excite single muscle fibres.

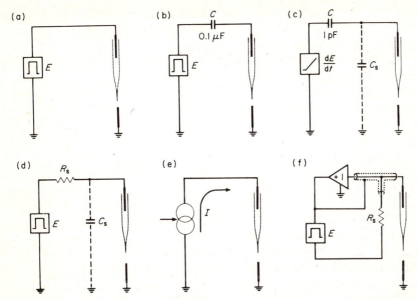

Fig. 31. Current sources. (a) Direct connection to voltage source. (b) Series capacitor. (c) Capacitor and voltage ramp. (d) Series resistor. (e) Controlled current pump. (f) Principle of Fein's (1966) current pump. Note the driven shield.

The *capacitor and voltage-ramp method* (Fig. 31(c)) is often used in the adjustment of negative capacitance and for measurement of microelectrode resistance (Chapter 4, Sections II.D and VII). At a time much greater than the time constant $R_{\mu E}(C + C_s)$, the current has the value $C dE/dt$ independent of microelectrode resistance. To keep the time constant small $C$ is usually made to be only 1–2 pF. A ramp of 500 or 1000 V s$^{-1}$ is then needed to get 1 nA to flow. It is impracticable to generate such ramps lasting longer than a few milliseconds, and so the circuit can only be used for short current pulses. Frank and Becker (1964) discuss a simple way of generating the voltage ramp. A 1–2 pF capacitor can be made by wrapping several turns of insulated wire around the wire that connects to the microelectrode and then adjusting the number of turns until the desired current is obtained.

The *series resistor method* (Fig. 31(d)) goes some way towards meeting most needs. Current pulses of any duration can be applied, and if the resistor $R_s$ is sufficiently large in value it will prevent gross leakage of current out of the cell between pulses; a value 20–50 times the input resistance of the cell is appropriate. If in addition we want the current during a pulse to be substantially independent of microelectrode resistance $R_{\mu E}$, we choose $R_s$ to be at least 20–50 times greater than $R_{\mu E}$.

Values of $10^8\ \Omega$, $10^9\ \Omega$, or $10^{10}\ \Omega$ are common. The current through the system is $E/(R_s + R_{\mu E} + R_{in})$, where $R_{in}$ is the input resistance of the cell. If we make $R_s$ too big, a large voltage will be needed to get only a few nanoamperes to flow. High value resistors are not especially easy to obtain, so when you find a source buy a lot of them. Resistors of $10^9\ \Omega$ and up need some care in handling; sticky fingerprints on the outside can reduce the resistance dramatically. If current pulses with short rise times are required, give thought to the physical placing of $R_s$. In order to minimize the stray capacitance $C_s$ shown in Fig. 31(d), $R_s$ should be as close as possible to the microelectrode.

An *active current pump*, symbolized in Fig. 31(e) as a controlled constant current source, can overcome all the difficulties of the methods given so far. It completely surpasses the series resistor method and should perhaps be considered the standard way of passing current through microelectrodes. A current pump is a "black box" equipped with an output terminal and some means of applying a command signal. When it receives a command it compels a current, linearly proportional to the command signal, to flow from the output terminal through any interposed electrical network to earth. The magnitude of the output current is governed only by the command signal and is *independent* (subject to practical limitations) *of the impedance connected to the output terminal*. Evidently a current pump is very like a constant-current stimulator. Stimulators are designed to pass currents measured in micro- or milliamperes through electrodes of resistance no more than about $100\ k\Omega$. For microelectrodes we need special circuits to pass currents measured in nanoamperes through resistances up to several hundred megohms.

The most ingenious current pump is that devised by Fein (1966). Fig. 31(f) shows its bare bones. In Fein's words, "analysis of this deceptively simple circuit" reveals that a constant current of magnitude $E/R_s$ flows in the output circuit, regardless of its resistance. The analysis is left as an exercise (Appendix I). In practical realizations of Fig. 31(f), $R_s$ is $10$–$100\ M\Omega$. Because of the difficulty of designing a voltage source which is both floating and precisely controllable, a somewhat more elaborate method is used to apply the command signal to $R_s$ (Appendix IV; see also Colburn and Schwartz, 1972).

Another kind of current pump is the Howland circuit used by Gage and Eisenberg (1969). Its operation (Exercise 21 of Appendix I) is described by Smith (1971). Lastly, there is what I call a *current clamp* (Appendix V) in which the circuitry measures the current flowing through a series resistor $R_s$ and forces it to be proportional to the command signal.

With appropriate design, a current pump can be provided with a driven shield (Fig. 31(f) and Appendix V). The driven shield eliminates the harmful effects of stray capacitance in a manner like that discussed in Chapter 4, Section II.C, and allows fast current pulses to be applied through cables up to a metre or so in length.

A final word about current pumps. Even black boxes have to obey the laws of physics, including Ohm's law. If you want to pass 250 nA through a 200 MΩ microelectrode, there will have to be 50 V across the electrode. The black box will have to apply rather more than that to its series resistor $R_s$. No circuit which operates from ±15 V supplies can possibly achieve this. If you must use such fierce pulses you need a high voltage current pump. Intended for ionophoresis, these are available commercially. Dreyer and Peper (1974) describe a very elegant circuit with two high voltage FET-operational amplifiers (priced at around £70 each). A much cheaper but less elegant version is given by Purves (1979).

### III. Current Monitors

There are four distinct ways of estimating the current through a microelectrode.

The *optimist's method* (Fig. 32(a)) uses the command signal to the current source as an indication of current flow. The advantages are simplicity and the ease with which very small (<0.2 nA) currents may be monitored. The disadvantage is that there is no true measurement; the microelectrode may in fact be completely blocked and pass no current at all, while the monitor continues happily to indicate a non-existent flow. The monitor outputs on many commercial ionophoretic current pumps and preamplifiers for current injection are of this type. It is very important to be aware of this so that you can be on your guard against electrode blockage. Monitoring of the voltage across the microelectrode is recommended.

A *galvanometer* connected in series with the current loop (Fig. 32(b)) makes a very accurate (but expensive, cumbersome and slow) current monitor.

*Ohm's law* allows the current to be calculated from the measured voltage drop across a known resistance in the current loop (Fig. 32(c)). The method of voltage measurement must not let any current escape from the loop; FET-operational amplifiers connected as voltage followers perform this function. The value of the resistance depends on the range of currents expected: 100 kΩ, 1 MΩ or 10 MΩ are common,

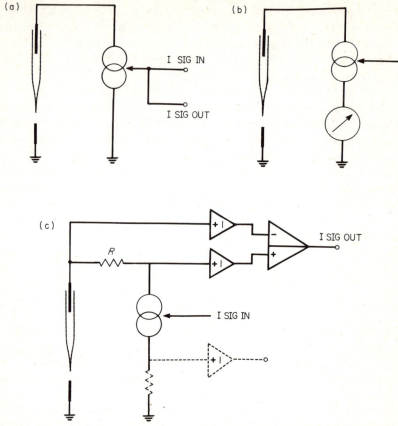

Fig. 32. Current monitors. (a) Optimist's method. (b) Galvanometer method. (c) Ohm's law method, showing two alternative positions for the current-sensing resistance $R$. The upper circuit requires two voltage followers and a differential amplifier. The lower circuit (interrupted lines) is simpler, but cannot be used if the current source has one terminal earthed (as is often the case with electronic current pumps).

giving a signal of 0.1, 1 and 10 mV nA$^{-1}$, respectively. The very cheap CA3140 operational amplifier is suitable for monitoring all but the smallest currents.

An *operational current-to-voltage* converter connected in series with the indifferent electrode (Fig. 33) has the special feature that it monitors the sum of all the currents injected into the preparation or extracellular medium. Its disadvantage is that it is highly susceptible to interference from alternating electric fields affecting the extracellular medium. Leads from stimulators and liquid-containing perfusion lines act as aerials, producing a hum current which is faithfully converted into a voltage

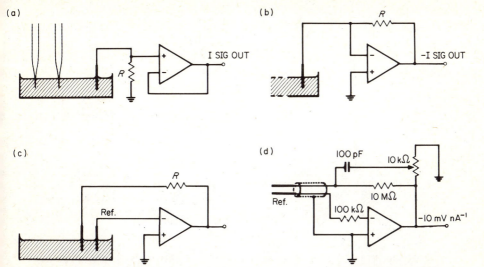

Fig. 33. Development of the current-to-voltage converter. (a) A variant of the Ohm's law method in Fig. 32(c). Passage of current alters the potential of the indifferent electrode, owing to the voltage drop across resistor $R$. (b) Simple current-to-voltage converter. Voltage drop across $R$ no longer appears at the indifferent electrode. (c) Advanced current-to-voltage converter, separating the two functions of an indifferent electrode (see Chapter 4, Section III.B). (d) Practical circuit using CA3140 operational amplifier. Bandwidth control (10 kΩ) reduces noise in low speed applications. The 100 kΩ resistor protects the amplifier from voltage overload. Note the shielding of the signal cables; this should be extended to cover as much of the preparation as possible.

signal by the monitor. The CA3140 operational amplifier is usually suitable, as in the previous method.

## IV. Alternate Current Injection and Potential Measurement

For intracellular current injection one often needs an indication that the current-passing microelectrode has penetrated a cell. Active current pumps can be provided with an output which, in the absence of applied current, will show the resting potential (see Appendix V). A similar result can be obtained by switching the lead from the microelectrode so that it connects alternately to a current source and a preamplifier (Fig. 34(a)). Some degradation of electrical performance is to be expected, since stray capacitance is increased. High speed relays have been used for fairly rapid alternation (Moore and Bloom, 1969). Electronic switching is discussed briefly in Section V.G below.

A "breakaway box" (Fig. 34(b)) is a multi-pole switch placed between

(a)     (b)

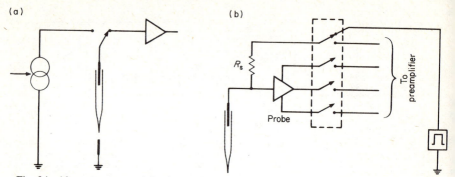

Fig. 34. Alternate current injection and potential measurement. (a) The microelectrode is switched to either a current source or a preamplifier. (b) A "breakaway" box isolates the probe from earth and from the rest of the preamplifier, allowing very large voltages to be applied to the microelectrode.

the probe and the rest of the preamplifier electronics. In the breakaway position large voltages can be applied to the microelectrode without risk of damaging the probe's input stage.

## V. Simultaneous Potential Measurement

The most natural method for simultaneous injection of current and measurement of membrane potential is the use of two microelectrodes. The required arrangements are sufficiently obvious that detailed description need not be given. It is best to use two indifferent electrodes, one acting as a current return and the other as a voltage reference, as explained in Chapter 4, Section III.B. The current monitor of Fig. 33(d) accomplishes this and gives a true measure of current flow.

The single microelectrode methods to be discussed here are inferior in nearly all respects to the two-electrode method. Their wide use results from the unfortunate fact that it is often difficult or impossible to impale one cell with two electrodes. Successful application of the single electrode technique demands an awareness of elementary electronics and circuit theory. The exercises of Appendix I should be consulted at this stage if they have not hitherto been attempted. Exercises 14–16 are especially relevant.

The essential problem is set out in Fig. 35. A current source controlled by a command signal I SIG IN passes current $I$ through a microelectrode of resistance $R_{\mu E}$ and an impaled cell whose passive electrical properties are represented by a parallel $RC$ network. The voltage seen by the preamplifier has two components, $E_1$ and $E_2$. $E_1$ is

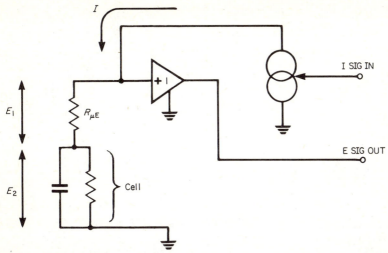

Fig. 35. Simultaneous current injection and potential measurement. The total signal $E$ seen by the preamplifier consists of a wanted part $E_2$ and an unwanted part $E_1 = IR_{\mu E}$.

the potential drop $IR_{\mu E}$ across the microelectrode, and $E_2$ represents the changes in membrane potential evoked by the passage of current. Clearly $E_1$ is of no biological import and must be removed somehow from the output voltage of the preamplifier. This may be achieved either by electronic subtraction or by direct cancellation.

### A. Current Pump with Electronic Subtraction

Figure 36(a) shows in simplified form the circuit used by the W-P Instruments range of current injection amplifiers, by Colburn and Schwartz (1972) and by the amplifier of Appendix IV. Disregarding for the moment capacitors $C_t$, $C_s$, $C_a$, $C_f$ and amplifier $A'$, we see that a differential amplifier subtracts a portion of the voltage at I SIG IN from the total voltage $E_1 + E_2$. The control labelled BAL allows the fraction of I SIG IN thus subtracted to be varied at will. Since the injection current $I$ is proportional to I SIG IN, it will be evident that the unwanted signal $E_1 = IR_{\mu E}$ is also proportional to I SIG IN. If the designer has done his sums right there will be some point of adjustment of BAL at which the subtracted signal is precisely equal to the unwanted signal.

### B. Current Pump with Direct Cancellation

Figure 36(b) shows the other way of removing the unwanted signal $IR_{\mu E}$ from the output. Again disregarding $C_t$, $C_s$, $C_a$, $C_f$ and amplifier $A'$, note

(a)

(b)

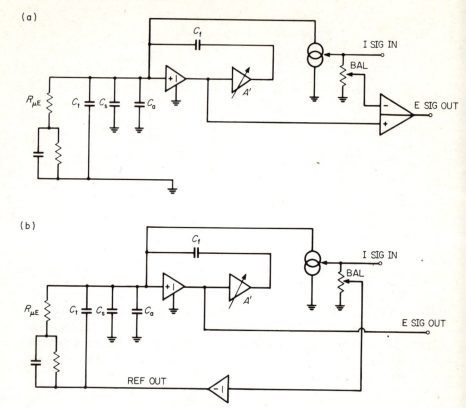

Fig. 36. Simultaneous current injection and potential measurement, showing the two methods for elimination of the potential across the microelectrode. (a) A signal proportional to the unwanted $E_1$ of Fig. 35 is subtracted from the total by the differential amplifier on the right. (b) A signal proportional to $E_1$ is inverted and applied to the preparation via REF OUT, where direct cancellation occurs.

that as in Fig. 36(a) the BAL control selects a portion of I SIG IN. Here, however, the balance signal is electronically inverted and applied to the return lead from the preparation. By the superposition theorem (Exercise 7 of Appendix I) the voltage seen by the preamplifier is the algebraic sum of $E_1$, $E_2$ and the inverted balance signal. Appropriate adjustment of BAL effectively cancels the unwanted signal.

## C. Old-fashioned Bridges

Before integrated operational amplifiers were available, current pumps of adequate performance could not be built, and so current sources by

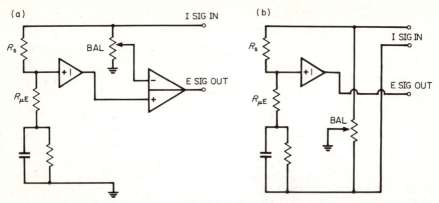

Fig. 37. Old-fashioned bridge circuits for simultaneous current injection and potential recording. Components relating to negative capacitance have been omitted. (a) Electronic subtraction of unwanted signal. (b) Direct cancellation method. Note that the voltage source supplying I SIG IN in (b) must be isolated from earth.

necessity consisted of a voltage source and series resistor. Circuits similar in conception to those just described were used for simultaneous current injection and potential recording (Araki and Otani, 1955; Frank and Fuortes, 1956; Bach-y-Rita and Ito, 1966; Martin and Pilar, 1963). The circuits were generally drawn to show their likeness to the bridge devised by Wheatstone for measurements of resistance. The unwanted signal $IR_{\mu E}$ was removed by one or other of the two methods noted above. The two types of bridge circuit are shown in Fig. 37, drawn to show their likeness to the modern circuits rather than to the Wheatstone bridge. Use of these circuits is declining, leaving only a historical legacy: the BAL control of Figs 36 and 37 is often called "bridge balance".

### D.  Comparison between Methods

From the brief description given so far there would seem little to choose between the methods of electronic subtraction (Section A) and direct cancellation (Section B). There are, however, real differences between them, and in my opinion electronic subtraction is decidely superior. It is worth inquiring into the reasons, since this topic has not been discussed previously.

First, the cancellation method monopolizes the indifferent electrode, whereas in the subtractive method it can be earthed directly or taken to a current-to-voltage converter, as the experimenter pleases. Second, the indifferent electrode in the cancellation method is not at a fixed potential but may undergo excursions of several volts. In the simplest experimental arrangement this is of no consequence, but if, for example, a second recording microelectrode is introduced, the changing reference potential produces complications. Again, the changing reference potential is likely to interfere with the correct operation of current sources used for ionophoresis. The subtractive method is free from these difficulties since the reference potential is held constant, usually at 0 V.

The third reason is more subtle. In Fig. 36(a) the three components of the total input capacitance ($C_t$, $C_s$ and $C_a$; see Chapter 3, Section V and Chapter 4, Section II.C) are in parallel as seen by the preamplifier and current source. Adjustment of the negative capacitance circuitry $C_f$ and $A'$ effectively removes them from consideration, and all is well. In Fig. 36(b), however, the transmural capacitance $C_t$ of the microelectrode bears a different relation to the REF OUT signal than do the other components $C_s$ and $C_a$; REF OUT sees $C_t$ in series and $C_s$ and $C_a$ in shunt. This asymmetry is closely related to that discussed in Chapter 4, Section II.D. Analysis shows that with full capacitance compensation applied the time derivative of REF OUT appears at the preamplifier input. Since the derivative of a step voltage change is an "impulse" of infinite amplitude, this means that the preamplifier's dynamic range will be exceeded by a spike at the start and end of a current pulse. The amplifier will remain saturated for a time depending on its overload characteristics but roughly proportional to the product $R_{\mu E}C_t$. The resulting loss of information can be severe, especially since one of the best methods for adjustment of bridge balance (Section E below) needs an accurate portrayal of events occurring in the first millisecond or so after a pulse is applied.

### E. How to Balance the Bridge

Decide whether accurate bridge balance is needed. If you only want to measure spike frequency of a neurone during current passage, it is best to leave the bridge frankly unbalanced, adjusting the BAL control merely to prevent the trace from disappearing off the oscilloscope screen.

Accurate balance is possible only if the microelectrode resistance is constant. Real microelectrodes are non-linear during current passage (Chapter 3, Section VII). Test the electrode before impalement to find the extent of its quasi-linear region, remembering that some electrodes behave better for outward current pulses and others behave better for inward pulses. The all too common behaviour of Fig. 38(a) and (b) excludes an electrode from accurate experimentation.

A primitive method is to balance the bridge with the microelectrode

(a)                                    (b)

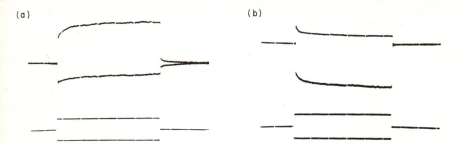

Fig. 38. Change of microelectrode resistance during current pulse. Superimposed traces showing effect of inward and outward current. Upper pair of each figure is the voltage record; lower pair is the current record. (a) Type I non-linearity, showing increase of resistance during outward current flow. (b) Type II non-linearity, showing fall of resistance during outward current flow. Calibrations are not given, since similar behaviour may be seen over a range of currents and sweep speeds, depending on the properties of the electrode under test. The traces show the time-dependent aspect of the steady state non-linearities discussed in Chapter 3, Section VII.

extracellular, then to impale and hope that $R_{\mu E}$ has not changed. This method nearly always gives the wrong answer, often grossly so (Schanne, Kawata, Schäfer and Lavallée, 1966; Schanne, 1969). The resistance may decrease during impalement owing to partial breakage of the tip. Usually, though, it increases because the conductivity of intracellular fluid is less than that of extracellular fluid.

The input resistance of most cells with action potentials falls to a low value at the peak of the spike; at this time the change in potential seen as a result of current injection is almost entirely due to $IR_{\mu E}$. Bridge balance can therefore be adjusted so that injection of current does not alter the displayed overshoot (Martin and Pilar, 1963). The method fails if the cell does not have a suitable action potential, or if for some other reason an action potential cannot be evoked. Severe inaccuracies result if the injection of current alters the true overshoot.

The best method is that popularized by Engel, Barcilon and Eisenberg (1972). In essence, the idea is to adjust the balance so that the displayed trace "looks right", i.e. so that the charging of the membrane capacitance at the onset of the pulse begins smoothly from the baseline potential. The unwanted component $IR_{\mu E}$ of the total potential is identifiable by its abrupt onset coincident with the applied current, whereas the wanted signal lags in time owing to the membrane capacitance. The method would be completely obvious and simple were it not for the stray capacitance of the input circuit. The effect of this is to distort the wanted signal and, much more importantly, to cause a lag in the time course of the injected current $I$. The latter effect results in a

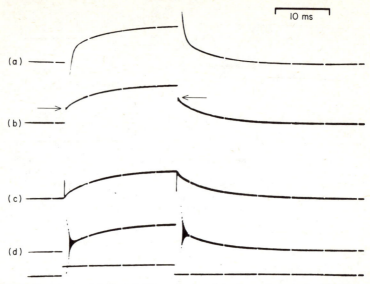

Fig. 39. Adjustment of bridge balance during injection of current. (a) No negative capacitance applied. (b) Correct application of negative capacitance reveals faulty bridge balance. Note the abrupt steps preceding the take-off points (arrowed) for the charging of membrane capacitance. (c) The bridge has been accurately balanced. (d) Too much negative capacitance.

disparity between the voltage $IR_{\mu E}$ and the balance signal; a transient artefact appears on the oscilloscope screen (Fig. 39(a)). Its time course is governed by the time constant $R_{\mu E}C_{tot}$. To allow accurate bridge balance this time constant must be much shorter than the cell's charging time. Careful adjustment of negative capacitance is nearly always needed to ensure that this condition is met (Fig. 39(b), (c)). The adjustment must be made while the responses are viewed at high sweep, say 0.5–5 ms per division. When full capacitance compensation has been applied the transient artefacts at the start and end of the current pulse are reduced to vertical spikes (Fig. 40). Any low-pass filter in the signal chain should be switched off to avoid the effect of Fig. 30(b).

## F. Errors and Artefacts

Most errors result from faulty bridge balance or from attempts to pass too much current through high resistance microelectrodes. An hour's practice with the analogue circuits of Fig. 41 will reveal any limitations of instrumentation or bridge-balancing technique. There is no simple way

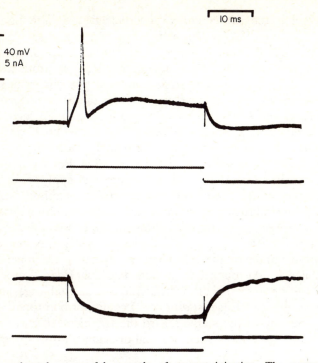

Fig. 40. A moderately successful example of current injection. The responses were obtained from a sympathetic neurone of a rat embryo, grown in cell culture for 4 days. Note the spiky transient at the start and end of the current pulses. A small degree of bridge imbalance is evident.

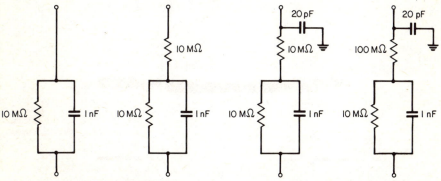

Fig. 41. Suggested test circuits for practising bridge balance, in increasing order of difficulty from left to right. The parallel combination of $R$ and $C$ (10 MΩ and 1 nF in each case) represents the passive electrical properties of an impaled cell.

of overcoming the limitations of microelectrodes, although bevelling (Chapter 2, Section VI) should help.

Current injected from a microelectrode has to traverse the cytoplasm as well as the cell membrane, and in so doing it produces potential gradients within the cell. The transmembrane potential changes evoked by current flow cannot therefore be the same at all parts of the membrane, except in the very special case of the microelectrode's being at the centre of a spherical cell. The consequent difficulties of interpretation, some of which are fortunately more hypothetical than real, have been discussed by Engel *et al.* (1972), Peskoff and Eisenberg (1975) and Purves (1976). These comments apply equally to the single-microelectrode and two-microelectrode methods of current injection with simultaneous potential measurement, although the precise effect of cytoplasmic current flow differs in the two methods.

The transient artefacts (Fig. 40) discussed in the previous section are an important guide to the adequacy of capacitance compensation and bridge balance. At one time it was fashionable to introduce extra circuitry to suppress the artefacts, with the inevitable result that the balancing method of Engel *et al.* (1972) could not be properly executed. This may be why many people regard the technique of current injection with suspicion. One often reads a ritual disclaimer to the effect that such

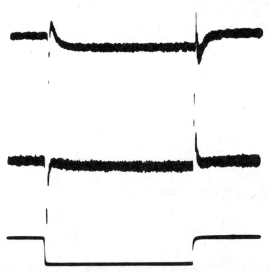

Fig. 42. Uncompensatable tip capacitance. Full capacitance compensation applied. Top trace: microelectrode immersed to a depth of 5 mm. Note that accurate bridge balance is impossible. Middle trace: microelectrode tip just in contact with bathing solution. Bottom trace: time course of applied current. The pulse was 5 ms in duration.

and such a result must be taken as approximate because it was obtained by a single-microelectrode method. A statement like this conceals one of two quite different circumstances. The experimenter may simply not have known how to adjust bridge balance; in this case the result is not "approximate" — it is wrong. Alternatively, it may genuinely have been difficult to locate the correct bridge balance; this occurs when high resistance microelectrodes are used on cells whose charging time constant is short. In favourable circumstances, as when the impaled cell has a high input resistance and long time constant, the single-microelectrode method need not be used apologetically for it is capable of high accuracy.

It will be recollected that a small part of the transmural capacitance $C_t$ of a microelectrode is "distributed", i.e. mixed up with the resistance of the tip region (Chapter 3, Section V). Microelectrodes of high resistance, immersed deeply within the bathing solution, may show transient artefacts that cannot be converted into vertical spikes by application of full capacitance compensation (Fig. 42). The resulting inaccuracies of bridge balance are usually rather small; the only practicable way to eliminate them is to reduce the depth of the bathing medium.

### G. Multiplex Method

Some of the difficulties associated with bridge balance and electrode non-linearity can be overcome by applying the injection current as a series of pulses, the membrane potential being measured only during the intervals between pulses. Electronic switching can be arranged to allow very rapid alternation between current injection and voltage measurement. If the measurement phase is sufficiently short with respect to the cell's input time constant, the membrane potential will not "droop" significantly, and the effect is one of continuous current injection and potential measurement. Bader, MacLeish and Schwartz (1979) and Merickel (1980) give circuits and discuss voltage-clamping with a single microelectrode (Wilson and Goldner, 1975). The multiplex method requires the time constant of the input circuit to be extremely small; the preamplifier must have well-designed negative capacitance circuitry.

# 6
# *Ionophoresis*

## I. Introduction

Ionophoresis is a convenient way of applying tiny quantities of drug or neurotransmitter in the neighbourhood of a receptive cell or of injecting inorganic ions or marker dyes into a cell. It is also a remarkably imprecise way. The absolute rate of release (and hence the concentration attained near the microelectrode tip) is usually unknown within a factor of 2–5. The speed with which drug concentrations at a cell surface can be changed at the onset of an ejecting pulse depends not only on distance from the microelectrode tip but also on the physical properties of the pipette itself. Ionophoresis can be used in rigorously quantitative studies only with the help of numerous checks and controls. Intending users of the method should not be put off by these discouraging remarks, for ionophoretic drug application is not at all difficult to do, and in many types of experiment the quantitative uncertainties are irrelevant.

Practical techniques and pitfalls of ionophoresis are discussed in detail by Curtis (1964), Krnjević (1971), Globus (1973), Bloom (1974), Kelly (1975), and Kelly, Simmonds and Straughan (1975). The reader is advised to read as many of the earlier texts as possible, since the present account makes no pretence to completeness.

## II. The Physics of Ionophoresis

### A. The Transport Processes

Four mechanisms contribute to or affect the release of drugs from micropipettes. They are *ionophoresis* itself, *diffusion*, bulk flow due to gradients of *hydrostatic pressure*, and bulk flow due to *electro-osmosis*. The disentangling of these four contributions has not been easy, especially since the interactions between them are more complicated than mere algebraic summation. The first theoretical study of note was that of Krnjević, Mitchell and Szerb (1963) who correctly identified the transport processes and gave quantitative expressions for the steady state release resulting from each process acting alone. Subsequent refinements have dealt with the interactions (Purves, 1977; Hill-Smith and Purves, 1978) and with the time course of release (Purves, 1979).

Ionophoresis is the electrophoretic migration of ions in a gradient of electric potential (Bockris and Reddy, 1970). If this were the only transport process at work, release of a given ion species would be governed by the relation

$$q = -nI/zF, \tag{11}$$

where $q$ is the efflux in moles per second, $I$ is the current, $z$ the charge number of the ion and $F$ is Faraday's constant (96 480 C mol$^{-1}$). The empirically determined parameter $n$ is the *transport number*, which expresses the fact that only part of the current is carried by ions of a given species. For example, during the ejection of a cationic drug by outward current flow, anions from the external medium are simultaneously transported backwards into the pipette. The minus sign in equation (11) arises from the fact that outward currents traverse the pipette of Fig. 6 in the negative $r$ direction. This sign convention is really needed only in the more advanced treatments which follow; it need not cause confusion in equation (11) if it is remembered that the efflux $q$ must be positive. Equation (11) has seen much service in the ionophoretic literature, where it is usually but wrongly called "Faraday's law". The name "Hittorf's law" would be more appropriate (Purves, 1979). If the ejecting current is applied in the form of a brief current pulse during which an electric charge $Q$ coulombs passes through the pipette, the quantity of drug released is $nQ/zF$ moles. The behaviour of real pipettes is often found to correspond reasonably well with Hittorf's law, provided that the ejecting current is not too small. Values of the transport number $n$ are found to be 0.1–0.5 for cations but sometimes much less for anions. The inadequacy of this simple theoretical framework becomes apparent if $I$ in equation (11) is set to zero, when the release $q$ should also become zero. Experimentally this is never observed. Instead a small spontaneous leakage persists. Worse still, if $I/z$ is made positive, equation (11) predicts a steady *influx* of drug, which is clearly impossible.

Diffusion is an expression of the thermal motion of mobile particles.

It may be regarded as the migration of particles, including ions, in a gradient of chemical potential (Bockris and Reddy, 1970). For a conical pipette (Fig. 6) of internal tip diameter $2a$ and included angle $2\theta$ radians containing a substance at initial concentration $C_0$, the steady state diffusional release is

$$q_D = \pi D C_0 \theta a, \tag{12}$$

where $D$ is the diffusion coefficient. If $2\theta = 8°$, $2a = 0.1\ \mu m$, $D = 1\ \mu m^2\ ms^{-1}$ and $C_0 = 1$ M, then $q_D$ is 11 fmol s$^{-1}$, equivalent (for transport number 0.5) to an ejecting current of 2.1 nA. If $2a = 0.4\ \mu m$ and $C_0 = 3$ M, the leakage is 130 fmol s$^{-1}$, equivalent to a current of 25 nA. Diffusional leakage can be a great nuisance, frequently being large enough to produce pharmacological effects. It can be made small by minimizing one or all of $C_0$, $\theta$ and $a$. As Krnjević et al. (1963) pointed out, this is more or less the same as making the microelectrode's resistance large. The second way to minimize the leakage is to apply a continuous "retaining current" in the opposite direction to the ejecting current.

Bulk flow through the tip is driven by a hydrostatic pressure gradient resulting from gravity, and is opposed by viscous resistance. The release due to this cause acting alone is

$$q_H = 3\pi\theta\rho g h C_0 a^3/8\eta, \tag{13}$$

where the gravitational attraction $g \approx 9.8$ J kg$^{-1}$, $\rho$ is the density of the filling solution, $\eta$ its viscosity and $h$ is the height of the liquid column. This expression is notable for its dependence on $a^3$, showing that the leakage due to bulk flow increases rapidly with increasing tip diameter. In the following illustrative calculations we shall take $\rho = 1000$ kg m$^{-3}$, $\eta = 0.001$ Pa s and the column height $h$ as 20 mm. If, as before, $2\theta = 8°$, $2a = 0.1\ \mu m$ and $C_0 = 1$ M $= 1000$ mol m$^{-3}$, the leakage $q_H$ is found to be 1.8 fmol s$^{-1}$, equivalent to a current of only 0.36 nA. But in a coarse-tipped pipette with $2a = 0.4\ \mu m$ and $C_0 = 3$ M, $q_H$ is 390 fmol s$^{-1}$, corresponding to a sizable ejecting current of 75 nA.

The fourth transport process is bulk flow due to electro-osmosis. Electro-osmosis is one of several phenomena collectively called "electrokinetic", the others (Shaw, 1970) being electrophoresis, streaming potentials and sedimentation potentials. Electro-osmosis is a manifestation of the electrical double layer at the interface between glass and water (Chapter 3, Section VI). The mobile part of the double layer is positively charged and therefore tends to migrate in an applied electric field in the same way that free ions migrate electrophoretically. In a microelectrode the mobile part of the double layer can be imagined as a conical tube that encloses the bulk phase electrolyte solution. Application of a positive voltage to the stem of the microelectrode causes the tube, together with its contents, to slide towards the tip. The sliding motion is opposed by viscous resistance within the double layer. Quantitative treatments of electro-osmosis in ionophoretic microelectrodes have been given by Krnjević *et al.* (1963) and by Hill-Smith and Purves (1978). Qualitatively it is obvious that electro-osmotic bulk flow must assist the ionophoretic ejection of cations, but diminish the ejection of anions. This may explain the rather small and variable apparent transport numbers measured for anionic drugs (see review by Kelly, 1975).

### B.  The Relation between Current and Drug Release

The inadequacies of the simple Hittorf relation (equation (11)) were noted in Section A above. A better description of the rate of release $q$ is

$$q = - \frac{I}{2zF\{1 - \exp[I/2zFq_{\mathrm{D}}]\}} \, , \tag{14}$$

which was derived with a slightly different notation by Purves (1977). The sign convention is again such that negative currents eject cations. The derivation of equation (14) relies on several assumptions, of which the most important are (1) that ionophoresis and diffusion are the only transport processes acting, (2) that all ions present have the same diffusion coefficient and absolute charge number, and (3) that the drug concentration $C_0$ in the filling solution is equal to the electrolyte concentration in the external medium.

Equation (14) is plotted as the curve $w = 1$ in Fig. 43 (the other curves and the meaning of $w$ are discussed later). Although we cannot expect this curve to reflect accurately the behaviour of real ionophoretic pipettes, it provides a qualitative guide. In the absence of applied current the rate of drug release is $q_{\mathrm{D}}$ of equation (12). The relation between current and efflux is non-linear for small currents, but becomes linear when sufficiently large ejecting currents are applied. It has been shown elsewhere (Purves, 1979) that the linear region corresponds roughly to the application of ejecting voltages greater than about 100 mV, regardless of tip diameter and resistance of the pipette. When the ejecting voltage is less than 100 mV the curvature of the relation is such that it becomes impracticable to use current as a measure of the steady state release. Currents passed in the opposite ("retaining" or "braking") direction diminish the spontaneous release but cannot

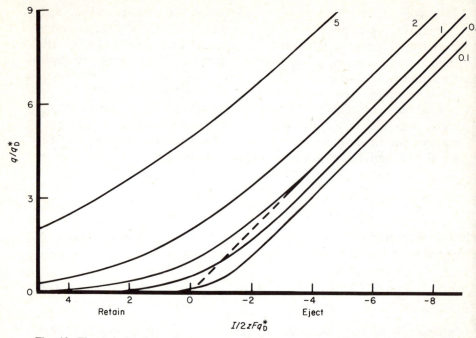

Fig. 43. Theoretical relation between current flow and drug release from ionophoretic pipette, calculated from equation (15). Numbers against curves are values of $w$. The interrupted line is the ideal relation $q = -I/2zF$.

suppress it entirely. In the example discussed here, a retaining voltage of 100 mV reduces the leakage to about 7% of its zero-current value; 200 mV reduces it to 0.3%.

An ideal ionophoretic pipette would attain its steady state rate of release the instant the current was turned on. Real pipettes may take from milliseconds to tens of seconds. Figure 44 shows some computed results for the time course of ejection (Purves, 1979) with the assumptions listed earlier in the text. Noteworthy features are that the onset of ejection is much slower than offset, and that prior application of retaining currents slows the onset still further and increases the discrepancy. This undesirable effect of retaining currents is due to withdrawal of drug from the tip region of the pipette, which has then to be refilled during a subsequent ejection phase. The time scale of release depends strongly on the geometry of the pipette, being several orders of magnitude faster for a fine-tipped pipette of large taper angle than for a coarse-tipped pipette of nearly cylindrical bore. This has some importance when ionophoretic drug application is used to investigate the speed of cellular responses.

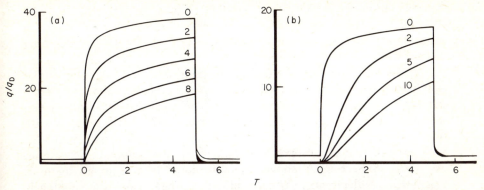

Fig. 44. Computed time course of ejection of drug from an ionophoretic pipette. The time scale is given by the dimensionless variable $T = Dt\theta^2/a^2$, and thus depends strongly on the geometry of the pipette's tip. The ejecting pulses were applied from $T = 0$ to $T = 5$. (a) Effect of various retaining currents on ejection during a pulse of fixed strength $I/2zFq_D = -40$. Numbers against curves are values of steady retaining current applied before and after the pulse. (b) Effect of retaining current applied at various times before an ejection pulse of fixed strength $I/2zFq_D = -20$. The retaining current of fixed strength $+20$ preceded the pulse by the times shown. Note that if the ejecting pulse were continued indefinitely, all the efflux curves in (a) would tend to the same value $q/q_D = 40.0$. Similarly in (b) all curves would tend to the value 20.0. The adverse effects of retaining current on the time course of ejection are clearly evident. (Reproduced from Purves (1979), by permission of Elsevier/North-Holland Biomedical Press.)

Drug solutions for ionophoresis are commonly used at concentrations anywhere in the range 5 mM–3 M, whereas the electrolyte concentration in the medium outside is normally fixed near 0.15 or 0.11 M. Assumption (3) above therefore appears unduly restrictive; we should like to know whether there is any theoretical reason to prefer a highly concentrated drug solution over a dilute one or *vice versa*. A generalization of equation (14) is derived in Appendix VI:

$$q = -\frac{I/2zF + k}{1 - \exp[-(1 + I/2zFk)\ln w]},\tag{15}$$

in which $k = q_D^*(1 - w)$, $q_D^*$ is the diffusional leak that the pipette would exhibit if its filling solution were at the same concentration as that $C_0^*$ of the external electrolyte medium, and $wC_0^*$ is the actual concentration of the filling solution. This result, which includes equation (14) as the limit $w \to 1$, is plotted in Fig. 43.

Two rather different lessons may be drawn from Fig. 43. The first is that if the range of ejecting currents to be used is of the same order as those shown, it matters a good deal what concentration of drug is chosen for the filling solution; values of $w \approx 1$ give the best approximation to the ideal straight line relation between current and drug release. The second, and more reassuring, lesson is that if one uses much larger ejecting currents than those of Fig. 43 the rate of drug release will be nearly independent of $w$, as suggested by the fact that all the curves have the correct slope at the right-hand end of the plot. The only remaining difficulty is to decide when a particular value of ejecting current is "large" in the sense just used. The argument will not be given in detail, since it follows that given by Purves (1979) for the special case $w = 1$. In essence, it involves the replacement of $q_D^*$ in equation (15) with a more readily observable parameter of the ionophoretic pipette. As noted in Section A above, the diffusional leak from a pipette is related to its electrical resistance, although when $w \neq 1$ the relation is complicated (see equation (7) of Chapter 3, Section IV). Pursual of this argument leads to the conclusion that for a range of $w$ between 0.5 and 10, the efflux of drug from a pipette of given geometry is independent of $w$ provided that the ejecting voltage exceeds 0.2–0.3 V. Thus, for example, if the minimum current to be ejected is 20 nA, it will not make much difference what drug concentration is chosen as long as the pipette resistance exceeds 10–15 M$\Omega$.

In practical terms, then, the user can choose a wide range of drug concentrations with little effect on his results, although the more dilute solutions have the advantage of needing less retaining current. There appears to be little point in choosing concentrations greater than 0.5–1 M; this merely wastes drug. Expensive or sparingly soluble drugs can be employed at much lower concentrations. It is probable, however, that the time course of release from very dilute solutions is slower than from concentrated solutions. Guyenet, Mroz, Aghajanian and Leeman (1979), in an experimental study of ionophoresis of substance P, suggested that the very slow ejection which they observed might be partly explained by the small concentration in the pipette (2–3 mM). More work on this matter is needed.

## III. The Practice of Ionophoresis

### A. Circuitry

Either the series resistor method (Chapter 5, Section II) or an electronic current pump can be used for passing current through ionophoretic pipettes. The current source must be able to supply a steady retaining current, adjustable in the range 0–25 nA or so, as well as the ejecting current itself. It is rarely necessary for the ejecting current to exceed

100 nA or the ejecting voltage actually applied to the pipette to exceed 10 V. In CNS work, ejecting pulses are normally many seconds or minutes in duration. Under these circumstances the current source need not have a spectacularly short rise-time; manual switching or control by relays is appropriate and may simplify the electronic design considerably. A high speed current pump has advantages only for the special instance of ultra-close application of drugs to rapidly responding receptors (Dreyer and Peper, 1974).

During the passage of large currents the excitability of neurones near the tip of the micropipette may be altered by the resulting extracellular potential changes as well as by pharmacological actions of the drug under test. The technique of "current balancing" overcomes this problem by ensuring that the algebraic sum of the currents flowing from a multi-barrelled assembly is zero. One of the barrels is chosen to carry the balancing current and is filled with some innocuous electrolyte (usually 1–2 M NaCl for extracellular ionophoresis). Figure 45(a) shows an arrangement that can be used when the current sources are isolated from earth. If the current sources are earth-referenced, as is usually the case for electronic current pumps, the arrangement of Fig. 45(c) offers an alternative. Circuits for isolated current pumps are given by Globus (1973) and Thomas (1975).

Willis, Myers and Carpenter (1977) described a circuit which was claimed to control electro-osmosis. The circuit is based on a misconception that electro-osmotic release is governed by the applied voltage, independently of the pipette's resistance. Since this premise is wrong, it is doubtful whether their "constant voltage" method of drug application has any advantage over the conventional constant current methods.

Tip blockage is an ever-present problem that promises hours of frustration for the unwary. It is signalled by an abrupt increase in the pipette's resistance. With the simpler circuits for ionophoresis there is no rapid way of measuring resistance and so blockage may go undetected. The best defence is the use of a current pump which monitors the voltage applied to the pipette. Less satisfactorily, blockage may be recognized by an abrupt decline in the current passed; this will only be possible if a proper current monitor is employed (Chapter 5, Section III). Very powerful current pumps can force large currents through badly blocked pipettes, but the passage of current is not always accompanied by the release of drug (Purves, 1979). The current pump of Appendix V has been designed with these considerations in mind. Beginners should avoid circuitry which fails to measure pipette current or voltage. Continuous monitoring of both is strongly recommended.

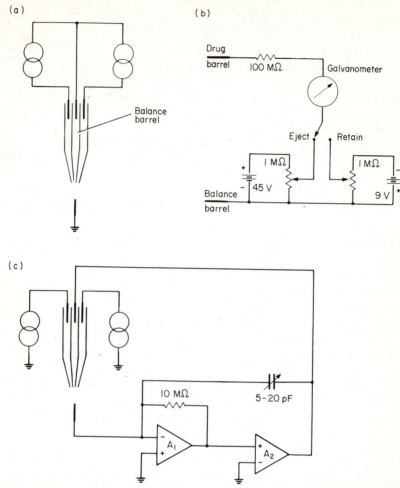

Fig. 45. Current balancing. (a) Isolated current sources pass current between the drug barrels and balance barrel of a multi-barrelled ionophoretic pipette. (b) A primitive isolated current source, with galvanometer to monitor the current. (c) Circuit to pass a balancing current equal to the sum of the drug currents but of opposite sign. $A_1$ is a current-to-voltage converter and $A_2$ is the feedback amplifier that clamps the current in the indifferent lead to zero.

## B. Intracellular Application

Intracellular injection (Chowdhury, 1969*b*) is most commonly made to alter the concentrations of inorganic ions (Coombs, Eccles and Fatt, 1955; R. C. Thomas, 1977) or to mark the cell for subsequent histological examination. Kater and Nicholson (1973) give a useful

(a)

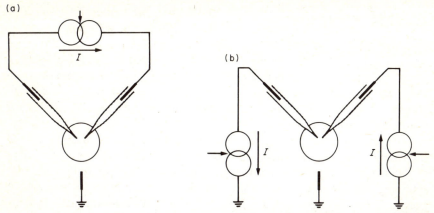

(b)

Fig. 46. Current balancing for intracellular ionophoresis. The arrangements (a) and (b) are exactly equivalent in their effects on the impaled cell, although this fact may not be intuitively obvious.

compendium of intracellular staining methods, but the two most popular markers for intracellular ionophoresis have been developed more recently. They are Lucifer Yellow (Stewart, 1978) and horseradish peroxidase (see discussion and references in Källström and Lindström, 1978; Neale, MacDonald and Nelson, 1978).

The magnitude of the ionophoretic current that can be injected intracellularly is restricted by the onset of membrane damage during large excursions of membrane potential. Much larger currents can be injected through one microelectrode if an equal and opposite current is withdrawn from a second electrode. This technique will be recognized as a variant of current balancing, which was introduced in the preceding section. Automatic current balancing has conventionally been achieved with a floating or isolated current source connected between the two microelectrodes (Fig. 46(a)), but the alternative method of Fig. 46(b) may be employed with some advantage. Here two identical current pumps are used, delivering currents of opposite sign, but controlled by the same command signal. The current pump described in Appendix V is suitable.

### C. Miscellaneous Hints

Cationic drugs are ejected more reliably than anionic drugs. In some cases it may be necessary to adjust the pH to produce optimal ionization (see Curtis, 1964) but many common drugs (acetylcholine, carbachol, atropine, hyoscine, noradrenaline, dopamine) can be used "straight from the bottle" and ejected as cations. If the glass fibre method is used to fill the pipettes, a few hundred microlitres of solution suffices.

Beginners are advised to use coarse or broken-tipped pipettes since these are less likely to block during current passage. Release of drug from a 2 $\mu$m pipette is much less precisely controllable than release from a 0.2 $\mu$m pipette, but in pilot experiments the important thing is to convince oneself that there *is* a release of drug. The details can be left until confidence in the method has been obtained.

Adjustment of retaining current remains an empirical matter, despite the theoretical considerations discussed above. Pipettes of very high resistance may require no retaining current at all, but in general the retaining current is to be adjusted until the preparation shows no signs of being affected by spontaneous leakage of drug. An initial guess may be made by setting the retaining current to a value that gives 0.1–0.3 V across the pipette. Excessive retaining current has disastrous effects on the speed of ejection during a subsequent current pulse.

## IV.  Other Methods of Drug Application

Until recently, ionophoresis was the fastest artificial method for applying drugs to cell surfaces, being surpassed in this respect only by presynaptic release of transmitter chemicals by nerve terminals. Lester and his colleagues, however, have introduced photochemical methods by which nicotinic acetylcholine receptors can be made to respond within the astonishingly short time of 10 $\mu$s after a stimulus. They developed photo-isomerizable compounds which in one configuration are inactive, but which can be rapidly converted to an active form by light pulses of appropriate wavelength (Lester *et al.*, 1980). Wider use of this method will depend on the synthesis of photo-isomerizable compounds capable of interacting with other kinds of receptor.

The main alternative to ionophoresis is pressure injection through micropipettes. The technique is discussed in some detail by McCaman, McKenna and Ono (1977), Källström and Lindström (1978), Sakai, Swartz and Woody (1979), and Spindler (1979). The advantage of pressure injection is that the quantities released are independent of the electrochemical properties of the active substance; thus it is readily possible to eject molecules having small diffusion coefficient or zero net charge or both. A heavy price must be paid for this advantage, since the rate of ejection is strongly dependent on the physical dimensions of the micropipette tip, being proportional to the third power of internal tip diameter. A given driving pressure may therefore produce widely differing flows in different pipettes.

# 7

# *Control of Vibration*

## I. Introduction

If one is fortunate enough to work on large skeletal muscle fibres in a basement laboratory, control of vibration may never be a problem. If one wishes to impale tissue-cultured neurones 15 $\mu$m in diameter while working in a first-floor laboratory with a wooden floor, elimination of mechanical transients may become a dominating passion. An engineering approach to the control of vibration would begin with a statement of what performance was needed, followed by measurements of the existing vibrations and attempts to discover and eliminate the causes. Local anti-vibration measures could then be designed. Physiologists, however, are not usually engineers and tend to omit the preliminaries. Even if you don't get around to measuring vibration you should at least try to work out whether it is mostly vertical or mostly lateral, and check that all doors in the vicinity are fitted with properly damped closing devices.

## II. Principles and Practice of Anti-vibration

To aid understanding of the principles of anti-vibration I have elected to give a simplified account of the physics involved, rather than to offer cookbook recipes or to describe any one method in detail. More practically-oriented accounts are given by Kopac (1964) and Wolbarsht (1964).

For convenience, only vertical motions will be considered, but it will be clear that lateral motions can be treated similarly. Figure 47(a) shows a stylized micromanipulator mounted on a rigid baseplate and

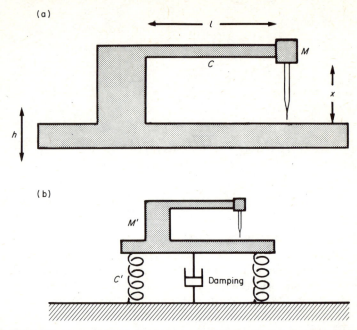

Fig. 47. Schematic arrangements considered in the treatment of mechanical vibration. (a) A micromanipulator supports a microelectrode and a mass $M$ on a horizontal arm of compliance $C$. (b) The manipulator and baseplate, of total mass $M'$, are in turn supported on flexible mountings of compliance $C'$.

supporting a microelectrode on the end of a probe arm. The mass $M$ of the probe is taken to be concentrated at the end, and the compliance measured at the end of the probe is $C$. (Compliance is the reciprocal of stiffness; if a force of 1 N applied to the probe produced a deflection of 1 mm, the compliance would be 1 mm N$^{-1}$.) Our aim is to minimize variations in the distance $x$ when the baseplate undergoes vertical excursions due, perhaps, to its ultimate connection with a wobbly floor.

In any practical system the resonant frequency of the probe assembly will be much higher than the vibration frequencies of the baseplate. Under these quasi-static conditions the displacement $\Delta x$ at the end of the probe is given by $\Delta x = \ddot{h}MC$, where $\ddot{h}$ is the instantaneous vertical acceleration of the baseplate. This result suggests that attention ought to be devoted more or less equally to reducing $\ddot{h}$, $M$ and $C$.

Reduction of $M$ is a common-sense matter: don't put heavy brass clamps or calomel electrodes at the end of the probe. Minimization of the compliance needs some slight knowledge of beam theory. The compliance measured at the free end of a cantilevered beam is $C = l^3/3EI$, where $l$ is length, $E$ is the Young's modulus of the substance from which the beam is made, and $I$ is the equivalent moment of inertia of the cross-section. For a rod of circular section, $I$ increases as the fourth power of the radius. Evidently the probe arm should be short and thick and made from a stiff material. However an arm made, say, from brass rod 1 cm in diameter will be heavy, and the mass of the arm itself will dominate the mass of the assembly. The same condition arises if we are successful in removing most of the mass $M$. The deflection at the end of a uniform beam of mass $m$ per unit length is $\Delta x = \ddot{h}ml^4/8EI$. This suggests that the arm should be made of a material with high specific stiffness (Young's modulus divided by density). Of the common materials, steel, aluminium, glass and (remarkably) wood have about the same specific stiffness. Brass is only about half as good, and carbon fibre composites are about four times better.

The micromanipulator has not so far entered the discussion, but it is clear that the minimum compliance of the assembly may be dictated by the compliance inherent in the manipulator. Practically then, the probe arm should not be made ultra-rigid at the expense of a greatly increased mass. As well as keeping the mass small, every attempt must be made *to keep the probe short*. An order of magnitude improvement should result from cutting the probe to half its length. The ultimate performance would be obtained from a stubby tubular carbon fibre probe arm, with the microelectrode clamped by a small nylon screw or tiny metal spring. If you choose to attach the microelectrode directly to a preamplifier probe (there are good reasons for this; see Chapter 4, Section II), you will be stuck with a large mass. There is not much you can do about this, but you can often reduce the compliance. Some commercial preamplifiers are supplied with a ridiculously flexible arm. Replace it with an aluminium one.

The anti-vibration measures discussed hitherto have the merit of involving only trivial cost and trouble. The same cannot be said for attempts to minimize the acceleration $\ddot{h}$. Figure 47(b) shows a baseplate of mass $M'$ supported with compliance $C'$. A full dynamical analysis will not be given here, but approximate analysis shows that for very brief vertical accelerations of the floor, $\ddot{h}$ is inversely proportional to $(M'C')^{1/2}$. Our aim is therefore to make both $M'$ and $C'$ as large as possible. An upper limit to $C'$ is set by the inconvenience of a very flexible mounting, the maximum tolerable compliance being perhaps $100 \, \mu\text{m N}^{-1}$. One has to increase the mass $M'$ until money, space, or floor strength gives out. A total mass of several hundred kilograms might be used, although 50–75 kg is more usual. Anything heavy can be piled on the baseplate, provided that it doesn't wobble or bend.

Compliant mountings can be made from metal springs, inflated bicycle or motor scooter tyres, rubber blocks, or tennis or squash balls. Commercial anti-vibration tables use compressed air which supports a rubber diaphragm on which the baseplate rests. The advantage of these commercial tables is not that they use air, but that they include a servo mechanism to stabilize the height of the baseplate. The feedback loop is slow to operate and does not reduce the short term compliance, but if an extra load is put on the table, air pressure under the diaphragms is eventually increased to compensate. Thus such a table can have a high short term compliance without the inconvenience mentioned earlier.

Any method of compliant mounting needs to have *damping* so that energy communicated to the baseplate can be dissipated. If there is too little friction in the mountings the baseplate will bounce up and down many times after a disturbance, possibly setting up a sloshing movement of the saline bathing the preparation. This is called "underdamped". If there is a lot of friction, movements of the floor can be passed on directly to the baseplate, and the benefits of compliant mounting are lost. This is "overdamped". We aim roughly for "critical damping", meaning that the system will recover from a disturbance as rapidly as possible without overshoot. Damping can be provided by mechanical friction as in the shock absorbers of old-fashioned cars, by fluid friction as in the shock absorbers of more recent cars, or by arranging that a part of the load is taken by foam rubber pads.

There is some theoretical advantage in splitting the total mass $M'$ and the total compliance $C'$ into two or more parts to give multi-stage isolation. The arrangement is cumbersome and bulky and not often used (but see Kopac, 1964).

A common design fault is insufficient rigidity of the baseplate. Vibration applied via the mountings $C'$ in Fig. 47(b) will then cause the baseplate to flex, with predictably bad results. A steel or cast iron plate at least 1.5 cm thick should be stiff enough if mounted intelligently. The baseplate sold with some micromanipulators (e.g. Leitz) is too flexible for delicate work, and may need to be stiffened with a steel bar bolted to the underside.

Finally, it should be mentioned that lateral vibrations are often more troublesome than vertical vibrations. Most commercial tables are notably poor at eliminating lateral accelerations. Rubber and metal spring mountings can be designed to have good lateral compliance, but the ultimate weapon is to suspend the whole affair on chains from the ceiling. Structural advice from a surveyor might prudently be sought beforehand.

# Appendix I: Electronics for Microelectrode Users

How much electronics should the microelectrode user know? Electronics covers a wide field and few electrophysiologists have the time or inclination to explore it fully. The essential things are to have a clear understanding of current flow and potential in the microelectrode circuit and to be able to use electrophysiological apparatus. It is far more important to be capable of operating an oscilloscope than to know how a transistor works.

This appendix outlines some areas of study for readers with little previous knowledge of electronics. Most of the material is given in the form of exercises that give practice in simple electrical calculations, illustrate important principles of electrophysiological measurement, or introduce interesting circuit theorems. Nearly all the exercises are easy, but total beginners may find it helpful initially to consult other texts (e.g. Donaldson, 1958; Kite, 1974). Numerical answers are given at the end.

## I. Ohm's Law and Other Circuit Theorems

Complete fluency in the application of Ohm's law is fundamental in all electrophysiological thinking.

*1*. In Fig. 48(a), calculate the current $I$ for the following pairs of values of $E$ and $R$: 1 V and 1 Ω; 5 V and 1 Ω; 1 mV and 1 MΩ; 5 mV and 1 MΩ; 50 mV and 5 MΩ. Note the advantage in the last three examples of expressing the current in nanoamperes. The units, millivolt, megohm and nanoampere form a "consistent set". Habitual use of them in electrophysiological thinking saves time because the powers of ten

$(10^{-3}$ V, $10^6$ Ω and $10^{-9}$ A) take care of themselves. The microelectrode user's versions of Ohm's law might be written:

$$I \text{ (in nA)} = E \text{ (in mV)}/R \text{ (in MΩ)},$$
$$E \text{ (in mV)} = I \text{ (in nA)} \times R \text{ (in MΩ)},$$
$$R \text{ (in MΩ)} = E \text{ (in mV)}/I \text{ (in nA)}.$$

2. In Fig. 48(b) a perfect constant current generator drives current $I = 2$ nA through resistor $R = 10$ MΩ. Calculate the potential $E$ across the resistor. Calculate $E$ in Fig. 48(c). Why are the two answers the same?

3. Calculate potential $E$ in Fig. 48(d) when $R = 10$ MΩ, 100 MΩ and 1000 MΩ. What is the relevance of these results for intracellular electrical recording, if $R$ represents the input resistance of the preamplifier?

4. Give the rule for combining resistances in parallel. Find $E$ in Fig. 48(e). Compare Fig. 48(e) with Fig. 48(b), noting that the only difference is the appearance of a 100 MΩ resistor in Fig. 48(e). This resistor is in parallel with the 10 MΩ; alternatively, it may be said to *shunt* the 10 MΩ resistor. In railway parlance, "shunting" is the diversion of traffic from a main line to a side branch. By obvious analogy a "shunt" in electronic parlance is an alternative pathway for current flow.

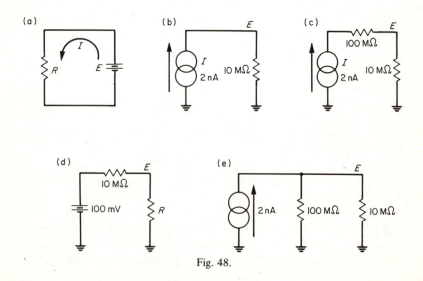

Fig. 48.

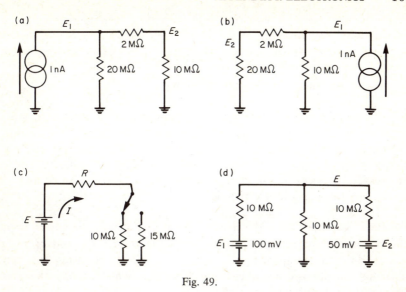

Fig. 49.

5. In Fig. 49(a) and (b) a constant current source supplies current to a resistive network. Find $E_1$ and $E_2$ in each case. The quantity $E_1/I$ is known as the *input resistance* or driving point resistance and is different for the two diagrams. The quantity $E_2/I$ is the *transfer resistance*. Note that it has the same value in both diagrams. This is a consequence of a very deep theorem in electric circuit theory: the reciprocity theorem. Among other things, the theorem asserts that in any network made of linear resistive elements, the position of a current source and a perfect voltmeter may be interchanged without affecting the voltmeter reading.

A physiological application of the reciprocity theorem is indicated in Exercise 8 below.

6. Calculate $I$ in Fig. 49(c) for the following pairs of values of $E$ and $R$: 100 mV and 0 Ω; 0.2 V and 10 MΩ; 1 V and 90 MΩ; 10 V and 1000 MΩ. In each case make the calculation for the switch in its left-hand and right-hand positions. Explain the relevance of your results for the controlled passage of current through microelectrodes, assuming the 10 MΩ and 15 MΩ resistors to represent the (not very constant) resistance of a microelectrode.

7. Calculate $E$ in Fig. 49(d). *Hint*: set $E_1 = 0$ and $E_2 = 50$ mV, calculate $E$ and record the result; then set $E_1 = 100$ mV and $E_2 = 0$; calculate the new $E$ and add it to the previous value to obtain the final

answer. Can you see why this trick is valid? The method illustrates another powerful theorem, the superposition theorem, which asserts that when a passive linear network is driven simultaneously by several sources, the total response at any part of the network is the algebraic sum of the responses to each source acting alone.

*8.* In many types of electrical recording, the shock artefact produced by nerve stimulation is sometimes so great as to obscure the response. One way to reduce the artefact is to arrange the stimulating electrodes so that they are more "efficient" in the sense that a smaller current can be passed between them and still excite the nerve. Show that the conditions for maximum efficiency are precisely those that would give the largest signal if the electrodes were used for extracellular recording of action potentials from the nerve. Hence account for the use of suction electrodes and paraffin oil pools for both stimulation and recording. *Hint*: apply the reciprocity theorem (Exercise 5), treating the action potential mechanism as a current generator which drives the action current through the nerve membrane.

This rather difficult exercise illustrates that electric circuit theory has relevance to the practical conduct of experiments. The reciprocity theorem deserves to be more widely known among electrophysiologists.

## II. Capacitance

At least some understanding of capacitance is necessary in most microelectrode work, especially when high speed recording or current injection is attempted (Chapter 4, Sections II.C and II.D; Chapter 5, Sections II and V.E). A firm grasp of the behaviour of real (positive) capacitance is of course absolutely necessary before any attempt is made to use the negative capacitance facility of modern preamplifiers (Chapter 4, Section II.D).

*9.* The usual definition of capacitance uses electrostatic concepts. If the two plates of the capacitor in Fig. 50(a) are initially at the same potential, and then a quantity of charge $Q$ coulombs is transferred from B to A, plate A will be found to be positive with voltage $E$ relative to plate B. The capacitance $C$ is defined as the ratio $Q/E$. Show that if at a given moment in time the potential across the capacitor is changing at a rate $dE/dt$, there is a current flow $I = C\,dE/dt$ "through" the capacitor. This view of capacitance, which relates a time-varying potential to current, is much more useful than the static definition. Whether

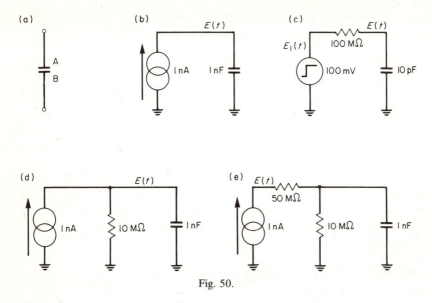

Fig. 50.

anything actually passes between the plates is a nice point debated in the higher reaches of electromagnetic theory; the important thing is the current flowing in the wires that connect to the plates.

*10*. In Fig. 50(b) a constant current source supplies a steady charging current to a capacitor. What shape is the waveform $E(t)$? Given that $E = 0$ at time $t = 0$, write down an expression for $E(t)$.

*11*. The term *impedance* is used in several places in this book, mostly in a qualitative way ("high impedance" or "low impedance"). It is important that the reader have at least a vague idea what it implies. Like resistance, impedance is a measure of the degree of obstruction to current flow, and is expressed quantitatively in ohms. In fact the impedance of a resistor *is* its resistance. But impedance has a more general meaning: the impedance presented to a signal may depend on the characteristics of the signal itself, whereas resistance is a property of the circuit alone. For example, the impedance offered by a capacitor to a steady potential is infinite; no current flows. On the other hand if the signal is varying rapidly in time, a large current flows "through" the capacitor and we must reckon the impedance to be small.

In Fig. 50(c) all voltages are initially zero, and then at time $t = 0$ the voltage of the source jumps instantaneously from 0 to 100 mV and remains there (this is called a *step voltage change*). Noting that the input signal changes infinitely rapidly at $t = 0$ and not at all thereafter, attempt to describe and draw the waveform $E(t)$ of the voltage across the capacitor.

*12*. In Fig. 50(c) a step voltage of 100 mV is applied at $t = 0$ as in Exercise 11. Calculate $E(t)$ as a function of time and plot it against $t$ with linear coordinates. Explain the relevance of your result for the recording of rapidly varying signals with microelectrodes, assuming $R$ to represent the microelectrode resistance.

*13*. Why is the product $RC$ called a *time constant*? Prove that the physical dimensions of resistance $\times$ capacitance are those of time. *Hint*: capacitance $\equiv$ current flow per unit rate of change of potential and resistance $\equiv$ potential per unit current flow. What are the units of ohm $\times$ farad and of megohm $\times$ picofarad?

*14*. In Fig. 50(d) all voltages and currents are initially zero. A 1 nA current step is applied at $t = 0$ to the $RC$ parallel combination. Calculate $E(t)$ and plot against $t$ with linear coordinates. What transformation could you make to get a straight line plot with semilogarithmic coordinates? Your graphs should show that $E$ reaches 63% of its final value at $t = 10$ ms. What is the value in milliseconds of the product $RC$? Explain the relevance of these findings for the measurement of passive electrical properties of cell membranes. Give the correct mathematical expression of which 63% is an approximation.

*15*. Referring again to Fig. 50(d), suppose that the current waveform consists of a rectangular pulse of amplitude 1 nA and duration $t' = 10$ ms. *Hint*: the rectangular current pulse can be regarded as the algebraic sum of two steps, the first being that of Exercise 14 and the second being a negative step (of $-1$ nA) applied at $t = 10$ ms. The sum of the two currents is zero for $t > 10$ ms. Calculate the response to the second step and add it algebraically to (subtract its absolute value from) the response already calculated in Exercise 14. The resulting waveform should look like a triangle with curved sides. The method of solution illustrates another aspect of the superposition theorem, in this case "superposition in the time domain".

Repeat the exercise for pulses of duration $t' = 20$, 50 and 100 ms. What are the implications for the measurement of input resistance and time constants of cells by injection of current pulses?

*16*. The arrangement of Fig. 50(e) is like that in Fig. 50(d) but includes a 50 MΩ resistor to represent a microelectrode through which current is injected. Without recalculating any results sketch the waveform $E(t)$. What is the effect of including the microelectrode resistance?

## III. Operational Amplifiers

The fundamental active device in modern analogue electronics is the integrated operational amplifier. Operational amplifiers have a complicated internal structure, but the user need not understand the inner workings at all. The reader of this book may have no intention of becoming involved in electronic design and construction, but may nevertheless find it useful to see how operational amplifiers are employed.

The following examples relate to an "ideal" operational amplifier

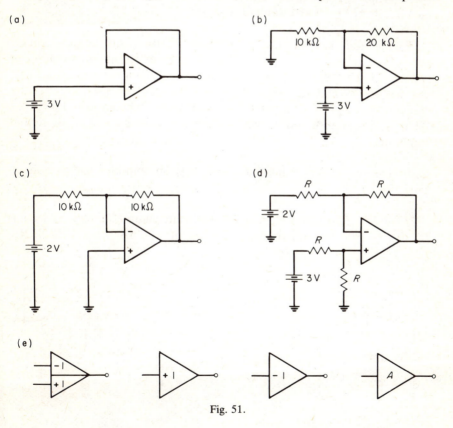

Fig. 51.

whose properties are listed in Exercise 17. Certain modifications to the circuits may be necessary when they are applied with real operational amplifiers. The CA3140 has been used in many of the circuits given in this book; its characteristics are so good that for most purposes it can be regarded as ideal.

*17.* The triangular symbol in Fig. 51(a) is an ideal operational amplifier having infinite gain, infinite input resistance, zero input leakage current, and zero output resistance. Find the output voltage. *Hint*: an operational amplifier adjusts its output voltage in such a way as to make the potential at the inverting input (−) equal to that at the non-inverting (+) input; this condition, together with the fact that no current can flow in or out of the input terminals, allows the output voltage to be determined.

*18.* After re-reading Exercise 17 carefully, find the output voltage of the circuits of Fig. 51(b), (c) and (d).

*19.* The four symbols in Fig. 51(e) stand for (from left to right) a differential amplifier, a voltage follower, an inverter, and an amplifier of gain *A*. Match them with the four operational amplifier circuits studied in Exercises 17 and 18. Given that a voltage follower does not alter the magnitude or sign of its input signal, what useful function could it serve? What is "impedance conversion"? What are some uses of a differential amplifier?

*20.* Design (and perhaps build and test) an amplifier with a voltage gain of +10.

*21.* Figure 52(a) shows a Howland current pump (Chapter 5, Section II). If $E$ is 100 mV, prove that a current of −20 nA flows through $R_{\mu E}$

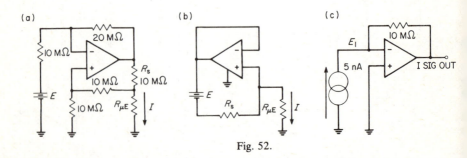

Fig. 52.

regardless of its value. A direct frontal attack on this problem leads to some complicated algebra. A more cunning approach is to find the output current when $R_{\mu E}$ has the value 0 $\Omega$. Next set $E = 0$ and replace $R_{\mu E}$ by a voltage source $E'$; show that no current is drawn from $E'$. These two results constitute the desired proof, although this fact may not be apparent at first sight.

22. Figure 52(b) shows Fein's current pump. Prove that a current of magnitude $E/R_s$ flows through $R_{\mu E}$ regardless of its value. *Hint*: find the potential between the two ends of $R_s$ and trace the pathway of the resulting current flow.

23. The circuit of Fig. 52(c) is a current-to-voltage converter. Find the output signal voltage. Point $E_1$ is known as a "virtual earth"; its potential is held very close to 0 V by feedback through the 10 M$\Omega$ resistor, so that the current source sees $E_1$ as though it really were earthed. This property makes the circuit useful as a current monitor (Chapter 5, Section III).

### Answers

1. 1 A, 5 A, 1 nA, 5 nA, 10 nA.
2. 20 mV.
3. 50 mV, 91 mV, 99 mV.
4. 18.2 mV.
5. (a) $E_1 = 7.5$ mV, $E_2 = 6.3$ mV; (b) $E_1 = 6.9$ mV, $E_2 = 6.3$ mV.
6. 10 nA, 6.7 nA; 10 nA, 8.0 nA; 10 nA, 9.5 nA; 9.9 nA, 9.9 nA.
7. 50 mV.
10. A steadily increasing voltage waveform (a "ramp"); $E$ (in mV) $= t$ (in ms).
12. $E(t) = 100(1 - \exp(-t/1 \text{ ms}))$ mV.
14. $E(t) = 10(1 - \exp(-t/10 \text{ ms}))$ mV;   $RC = 10$ ms;   63% $\approx 1 - 1/e$. Plot $E(t) - E(\infty) = E(t) - 10$ mV on a log scale against $t$ on the linear scale. Riggs (1963) gives useful hints on graph-plotting.
17. +3 V.
18. +9 V, −2 V, +1 V.
23. −50 mV.

# Appendix II: Current—Voltage Relation of Microelectrodes

Refer to Chapter 3, Section IV for a quantitative treatment of the linear resistive properties of a microelectrode, and to Chapter 3, Section VII for a qualitative description of the non-linear properties analysed below.

### Type I Non-linearity

The analysis will be based on three assumptions. (1) The pipette geometry is as shown in Fig. 6 and discussed in Chapter 3, Section IV. (2) The electrode is filled with an electrolyte solution of initial concentration $C_0$; the cations and anions have the same diffusion coefficient $D$ and absolute charge number $|z|$. (3) An infinitely thin "cation plug" occludes the lumen at the tip $r = r_0$ and prevents movement of anions into or out of the pipette. The plug itself has negligible resistance.

Since the anion flux is zero at $r = r_0$, in the steady state it must be zero for all values of $r$ within the pipette. The net flux of anions in the $r$ direction is given by the Nernst – Planck equation (see Purves, 1977, 1979) as the sum of electrophoretic and diffusional components:

$$j = -\frac{I}{2\,|z|\,F} - \pi\theta^2 r^2 D\,\frac{dC}{dr} = 0, \tag{16}$$

where $I$ is the current through the pipette in the positive $r$ direction and $C = C(r)$ is the anion concentration. Equation (16) is to be solved for $C$ subject to the boundary condition $C \to C_0$ as $r \to \infty$:

$$C = C_0\left(1 + \frac{I}{2\,|z|\,FDC_0\pi\theta^2 r}\right). \tag{17}$$

By electroneutrality this must also be the cation concentration. Substitution of equation (17) into the Nernst – Einstein relation (equation (5)) and thence into equation (2) gives

$$E = -\frac{RT}{|z|F}\frac{I}{I_{max}}\int_{r_0}^{\infty}\frac{r_0\,dr}{(r^2 + rr_0 I/I_{max})}$$

$$= -\frac{RT}{|z|F}\ln\left[1 + \frac{I}{I_{max}}\right], \tag{18}$$

where $I_{max} = 2|z|FDC_0\pi\theta^2 r_0$ is a limiting value of the outward current which can be passed through the pipette.

In addition to the resistive voltage drop given by equation (18), there will be a Nernst potential across the cation plug, of magnitude

$$\frac{RT}{|z|F}\ln\left[\frac{C_0^*}{C(r_0)}\right],$$

where $C_0^*$ is the cation concentration outside the tip. The total potential is

$$E' = -\frac{2RT}{|z|F}\ln\left[1 + \frac{I}{I_{max}}\right] + \frac{RT}{|z|F}\ln\left[\frac{C_0^*}{C_0}\right] \tag{19}$$

in which the second term represents the tip potential in the absence of applied current. The meaning of these results becomes clearer on introduction of the slope resistance at the origin, which will be called $R_{\mu E}$:

$$R_{\mu E} = \left.\frac{dE'}{dI}\right|_{I=0} = \frac{2RT}{|z|FI_{max}} \approx \frac{52\text{ mV}}{I_{max}}. \tag{20}$$

$I_{max}$ is now seen to be the outward current that would produce 52 mV across the nominal resistance $R_{\mu E}$. For example, a 2 M$\Omega$ electrode should pass no more than 26 nA, and a 10 M$\Omega$ electrode no more than 5.2 nA in the outward direction. Equation (19) can be written in the form

$$E' = -52\ln\left[1 + \frac{IR_{\mu E}}{52\text{ mV}}\right]\text{ mV} + \text{Tip potential}, \tag{21}$$

which is plotted in Fig. 9(c). Note that outward currents correspond to negative values of $I$.

### Type II Non-linearity

The analysis is based on four assumptions. (1) The pipette geometry is as shown in Fig. 6 and discussed in Chapter 3, Section IV. (2) The electrolyte solution filling the microelectrode is initially at concentration $C_0$, and the external electrolyte solution is at concentration $C_0^*$. (3) All ions have the same diffusion coefficient $D$ and absolute charge number $|z|$. (4) The passage of current produces an electro-osmotic bulk flow through the microelectrode, outward currents leading to outward flow.

Firth and DeFelice (1971) have given a formula for the resistance of an electrode through which a bulk flow $V$ is passing. With some changes of notation their result can be written

$$E = IR_{\mu E} \frac{(w - 1)(e^g - 1)(1 + (\ln w)/g)}{\ln w(we^g - 1)}, \tag{22}$$

where $R_{\mu E}$ is the nominal resistance (equation (7)) of the microelectrode in the absence of bulk flow, $w = C_0/C_0^*$ and

$$g = -V/\pi\theta^2 Dr_0. \tag{23}$$

The sign convention is such that $V$ takes negative values for outward flow.

The electro-osmotic flow (Krnjević et al., 1963; Hill-Smith and Purves, 1978) is

$$V = \varepsilon\xi I/\eta\sigma, \tag{24}$$

in which $\varepsilon$ is the permittivity of the filling solution, $\eta$ is its viscosity, and $\xi$ is the zeta potential. This expression is strictly applicable only if the conductivity $\sigma$ of the filling solution is constant along the pipette. In the case treated here $\sigma$ is not constant, and a full analysis of electro-osmosis would be very arduous, as Snell (1969) has indicated. Nevertheless, for a given current $I$ the electro-osmotic flow $V$ must have some definite value corresponding to a conductivity intermediate between that of the initial solution and that of the external medium. A crude way of expressing this is to take the geometric mean of the two conductivities, a procedure justified only by expediency:

$$\sigma = 2z^2F^2DC_0^*\sqrt{w}/RT. \tag{25}$$

Inserting equation (25) into equation (24) and thence into equation

(23), and making use of equation (7), we find

$$g = - IR_{\mu E} \frac{\varepsilon \xi (w - 1)}{\eta D \sqrt{w} \ln w} , \tag{26}$$

which now replaces equation (23).

The plot in Fig. 9(d) has been made with the following parameter values: $\varepsilon = 80 \varepsilon_0$, $\eta = 0.001$ Pa s, $\xi = 50$ mV, $D = 1$ $\mu m^2$ ms$^{-1}$.

# Appendix III: A Simple Preamplifier

Every completed electronic design represents a compromise between the conflicting requirements of performance, price, ease of use and ease of construction. This appendix describes one of many possible designs for a simple preamplifier that is suitable for student use and other not-too-demanding applications. The design procedure begins with the drawing up of a list of the main requirements. In this case we choose:

(1) Cost (at 1980 prices) to be less than £15, excluding power supply.
(2) Unity gain, independent of circuit adjustments.
(3) Negative capacitance.
(4) Push-button measurement of microelectrode resistance.
(5) Working input voltage range of at least ±1 V.
(6) Offset voltage adjustment of ±300 mV.
(7) Leakage current less than 15 pA.
(8) Rise time less than 100 $\mu$s with 20 M$\Omega$ source and no added capacitance.
(9) Thermal drift less than 50 $\mu$V K$^{-1}$.

For ease of construction we decide that the preamplifier will not have a separate probe, but will be coupled to the microelectrode by a short length of screened cable incorporating a driven shield. Also, we decide that there should be no internal batteries, since their eventual exhaustion would pose a difficult diagnostic problem to a student user.

The first crude design is shown in Fig. 53. The CA3140 operational amplifier will be used throughout because it is cheap and has sufficiently small leakage current (less than 10 pA) and thermal drift less than 10 $\mu$V K$^{-1}$. A3 is connected as a voltage follower with the driven shield joined to its output. A1 is a non-inverting amplifier whose gain is set by the ratio RV1/R2; a variable proportion of its output voltage is tapped

120

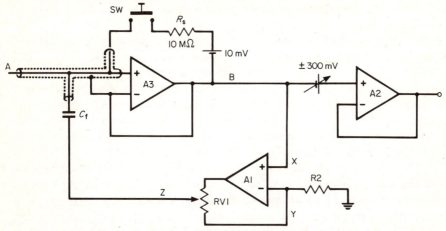

Fig. 53. First attempt at a design for a simple preamplifier.

off at point Z and applied via $C_f$ to the input to give negative capacitance. The 10 MΩ resistor and 10 mV "battery" will be recognized as forming a 1 nA current pump (Exercise 22 of Appendix I) to allow rapid measurement of microelectrode resistance when switch Sw is closed. Voltage follower A2 has been added merely because we suspect that the eventual form of the ±300 mV offset control will need to see a high impedance load. A2 will therefore function as a "buffer" to isolate the preamplifier circuitry from external loading conditions at the output.

A small simplification can be made immediately by omitting $C_f$ and driving the shield of the input lead from point Z. $C_f$ is thus effectively replaced by the core-to-screen capacitance of the input cable. Next we turn our attention to the operational amplifiers. Do we really need three of them? With a little ingenuity it will be found possible to make A1 do the job of A3 as well as providing negative capacitance. A3 can then be omitted. Consider points X and Y in Fig. 53. The operational amplifier keeps the potential at point Y equal to that at X by feedback through RV1. This suggests the use of X as the input to the preamplifier, and Y as the unity-gain output passing to the rest of the circuitry. We shall have to be careful what components are connected to Y, because the impedance to earth at this point determines the gain of the capacitance compensation signal; in Fig. 53 the gain of A1 is (1 + RV1/R2). This idea looks promising, but before we settle the details (such as the value of RV1) the "batteries" have to be eliminated since it has already been decided that real, physical, batteries should not be used.

There are several electronic devices that can maintain a fairly constant voltage between two terminals when supplied with a current of a few milliamperes. Ordinary silicon rectifier diodes can provide ~600 mV, and *zener diodes* are available in a wide range of voltage ratings. But these devices are apt to have excessively large temperature coefficients, and zener diodes generate much voltage noise. The ZN423 integrated circuit appears more suitable; it behaves as a precision voltage reference diode, giving about 1.2 V with very small noise and thermal drift. With the addition of a few resistors as voltage dividers, it will be easy to obtain the required 10 mV and ±300 mV.

The completed circuit design is given in Fig. 54. The 100 kΩ resistor in the input lead protects A1 against input voltages up to ±100 V. RV2 is a preset control which adjusts the offset voltage of A1 to zero (real operational amplifiers, unlike ideal ones, have an offset voltage of a few millivolts or tens of millivolts). D1 and D2 are the voltage reference diodes, and are bypassed with 0.1 μF capacitors for low noise. The non-inverting gain of A1 is governed by the ratio of RV1 to the parallel combination of the two 8.2 kΩ resistors. These resistors are effectively connected between point B and earth, from the standpoint of variations of signal voltage, since D1 and D2 present a very low resistance to signal currents, and the power supply terminals represent low impedance sources. Thus A1 has a gain of about 2. The amplified signal at the output of A1 is used only for negative capacitance. The signal at B, which is accurately at unity gain, passes to the offset control RV4 and output buffer A2. A low-pass filter consisting of the 1 nF capacitor and preceding resistors gives a fixed cut-off frequency of about 10 kHz.

The component values shown in Fig. 54 allow operation from a ±15 V supply. Parenthesized values permit the use of ±6 V as an alternative. The supply current is less than 20 mA.

### Construction

The input lead is 5–20 cm of shielded cable. The total core-to-screen capacitance should not exceed 15–20 pF. Ordinary screened wire for audio applications has a capacitance of about 300 pF m$^{-1}$, but special cables are fairly readily available with values down to 100 pF m$^{-1}$. The cable should be flexible; this rules out the use of most radio-frequency coaxial cables. Connection of the input lead to the non-inverting input of A1 should be made in mid-air or on PTFE insulators. This lead must *not* be run on a printed circuit board. The 100 kΩ resistor should be soldered directly to the input of A1, without an intermediate socket.

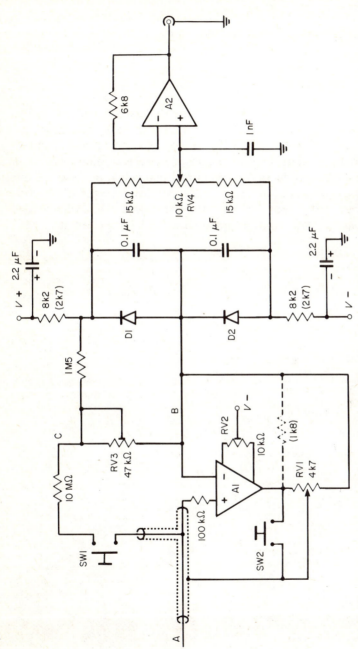

Fig. 54. Completed circuit design for simple preamplifier. Operational amplifiers are CA3140; diodes D1 and D2 are ZN423. Resistor values shown in parentheses are for operation from ±6 V power supply. RV1 and RV4 are front panel controls for adjustment of negative capacitance and offset voltage respectively.

The input lead must be screened as shown in Fig. 54, the screening being continued as far as the switch and the 100 kΩ resistor. The entire circuit should be enclosed in an earthed metal box.

### Setting Up

Earth the input (point A) and monitor the potential at B on a digital voltmeter or oscilloscope. Adjust RV2 for null (less than 1 mV). Connect a 10 MΩ resistor between A and earth, and monitor the preamplifier's output on an oscilloscope. Check that full clockwise rotation of RV1 produces oscillation and that full anti-clockwise rotation produces a quiescent output. If necessary reverse the connections to RV1. Note the output voltage and then close Sw. Adjust RV3 for approximately 10 mV shift in potential when Sw is operated.

# Appendix IV: Preamplifier with Current Injection

This circuit illustrates the level of performance that can be achieved by the amateur designer using readily available components. The specifications are similar to those of many commercial preamplifiers costing five times as much. Cost of the components, excluding power supply, is about £60 at 1980 prices.

### Specifications

Gain: adjustable to 1.000.
Input resistance: adjustable to at least 10 GΩ.
Leakage current: adjustable to 10 pA or less.
Working input voltage range: ±2 V on transients with maximum negative capacitance applied; ±10 V in steady state.
Absolute maximum input voltage: ±25 V from low impedance source.
Rise time: less than 5 $\mu$s with zero source impedance; less than 30 $\mu$s with 20 MΩ source shunted by 10 pF to earth (negative capacitance applied for optimal response without overshoot).
Negative capacitance: 12 pF can be compensated. This value can be increased by a simple modification to the probe.
Offset adjustment range: coarse ±300 mV; fine ±30 mV.
Current injection: up to ±12 nA d.c.; ±10 or ±100 nA pulsed. Maximum output current is 10 V ÷ ($R_{\mu E}$ + 100 MΩ). Analogue and TTL 5 V logic pulse inputs are available.
Current monitor: true measurement; scale factor 10 mV = 1 nA.
Noise: approximately 20 $\mu$V r.m.s. with zero source impedance and 10 kHz bandwidth; 65 $\mu$V with 20 MΩ source and negative

capacitance adjusted to give 10 kHz bandwidth (the Johnson noise of the source alone is 56 $\mu$V).

Drift: less than 50 $\mu$V K$^{-1}$.

## Circuit Description

As indicated in Fig. 55(a)–(d) the preamplifier has a remote probe connected by 0.5–0.6 m of 4-core-and-screen cable to the main chassis.

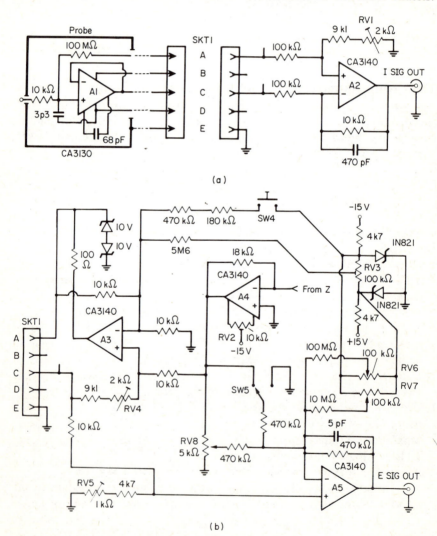

(a)

(b)

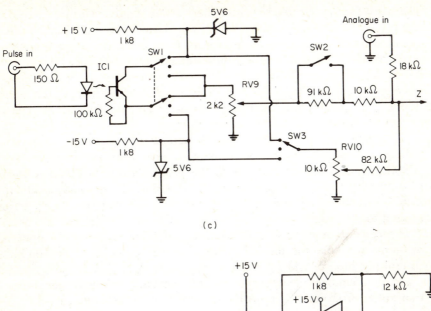

(c)

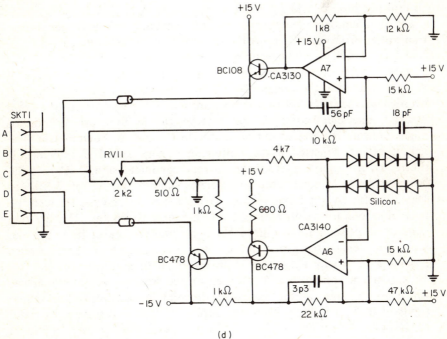

(d)

Fig. 55. Preamplifier with current injection. (a) Probe and current monitor. (b) Feedback and output board. (c) Current control board. (d) Negative capacitance board.

There are four circuit boards (excluding power supply): current monitor, feedback and output, current control and negative capacitance. The circuit is partly based on a design for a simple but very fast preamplifier given by M. V. Thomas (1977); those intending to construct the present design should consult Thomas' paper for a discussion of the novel method of applying negative capacitance via the supply lines to the input device. The preamplifier requires a stable, regulated, noise-free supply of ±15 V, from which it draws about 70 mA.

The probe uses a CA3130 BiMOS operational amplifier A1 in the voltage follower configuration. Current injection is achieved via the 100 MΩ resistor and terminal A of the probe socket. A 10 kΩ resistor provides limited protection against overloads. It is inadvisable to increase its value because it is in series with the negative capacitance applied to A1.

The current monitor is a differential amplifier A2 whose output is one-tenth the potential difference between the two ends of the 100 MΩ current injection resistor.

On the feedback and output board, current command signals are transmitted from point Z by A4 and summed with the unity-gain voltage signal by A3. This bootstrapping technique ensures that the 100 MΩ current injection resistor does not shunt the probe input. The output of A3 is restricted in the range ±10.6 V by a zener diode clamp to prevent overdriving the probe when attempts are made to pass current through a blocked microelectrode. Without this restriction the current monitor might indicate a non-existent current (see Chapter 5, Section III). The output of A5 consists of the unity-gain voltage signal from which is subtracted a variable part of the current command signal. RV8 is the bridge balance control, and Sw5 selects balance ranges of 0–100 or 100–200 MΩ. Sw4 is a push-button giving a current of 1 nA through the microelectrode for measurement of its resistance. RV6 and RV7 are fine and coarse offset adjustments. RV3 is an internal preset by which the leakage current at the probe can be set to zero. The 1N821 zener diodes provide ±6.2 V with very small temperature coefficient.

The current control board assembles three separate command signals which are summed at the virtual earth point Z. Rectangular pulses at standard TTL 5 V logic level are applied to an optoisolator to avoid earth loops. An output of either polarity is taken via Swl and amplitude control RV9. Sw2 selects 0–10 and 0–100 nA ranges. Direct current up to ±12 nA is controlled by Sw3 and RV10. Finally, an analogue input allows current signals of any waveform and polarity to be applied, the scale factor being 1 V = 10 nA. The 100 kΩ resistor between base and emitter of the phototransistor equalizes its turn-on and turn-off times.

The negative capacitance board provides the supply voltages for the probe. Under zero-signal conditions, terminals B and D are at ±7 V. These terminals are driven by the signal at C with a gain of about 0.7 to give a wide range of working input voltage at the probe, a feature which permits the preamplifier to pass appreciable current through high-resistance (e.g. dye-filled) microelectrodes. Negative capacitance is obtained with additional drive to terminal D adjusted by front panel control RV11. Emitter followers isolate the outputs from the capacitance of the probe cable. Ferrite beads on the emitter leads discourage parasitic oscillation. Additional stabilisation, if needed, can be secured by a 100–220 pF capacitor in parallel with the 1 k8 resistor and by increasing the 3p3 capacitor to 5p6 or 6p8. The ±15 V supplies to this board should be bypassed to earth with 22 $\mu$F, 22 V tantalum capacitors (not shown). A "zap" control (Chapter 1, Section II.G) has not been included on the circuit diagram. If this facility is wanted, a push button switch Sw6 should be added to connect terminal C to the wiper of RV11.

### Front Panel Controls

| | |
|---|---|
| RV6 | offset fine |
| RV7 | offset coarse |
| RV8 | bridge balance |
| RV9 | pulse amplitude |
| RV10 | d.c. amplitude |
| RV11 | negative capacitance |
| Sw1 | pulse polarity (double pole changeover, centre off) |
| Sw2 | 0–10 nA, 0–100 nA (single pole, single throw) |
| Sw3 | d.c. polarity (single pole changeover, centre off) |
| Sw4 | microelectrode test (normally open, push-to-close) |
| Sw5 | 0–100 M$\Omega$, 100–200 M$\Omega$ (single pole changeover) |
| Sw6 | "zap" (normally open, push-to-close; not shown in circuit diagram) |

### Construction

The use of 2% tolerance metal oxide resistors is recommended for all values less than 1 M$\Omega$. The 3.3 pF capacitor in the probe is a silvered mica type. Preset variable resistors should be multi-turn cermet types. The CA3140 operational amplifiers A2–A6 are supplied with ±15 V (not shown in circuit diagram). The ±15 V supplies to the feedback and output and negative capacitance boards should be bypassed to earth with tantalum capacitors (e.g. 22 $\mu$F, 22 V). Zener diodes marked only with a voltage rating are type BZY88.

The only critical parts of the circuitry are the probe and the negative capacitance board. With some care it will be found possible to solder the various components and leads in the probe directly to the pins of A1. The CA3130 operational amplifier is available in a TO-5 can or in a plastic dual-in-line package. Construction of the probe is easier with the former type, but both have been used successfully. M. V. Thomas (1977) suggests one particular physical design of the probe assembly, but this can be varied to suit personal preference. The method of mounting should ensure that the earthed metal case cannot contact any part of the electronics. The contents of the probe should be thoroughly cleaned and degreased, rinsed in distilled water, and allowed to dry completely before insertion into the case. If it should prove desirable to increase the range of negative capacitance, the 3.3 pF capacitor in the probe can be increased to 5 or 8 pF with slight degradation of the rise time. The larger values should be connected to the left end of the 10 k$\Omega$ protection resistor, not to the right end as shown in the diagram.

The circuitry on the negative capacitance board has gain at frequencies of several megahertz. Compact construction is advisable. In particular, the leads to the non-inverting input of A6 must be kept short.

### Setting up

(1) Turn off the current control switches Sw1 and Sw2. Monitor the output of A4 and adjust RV2 until the measured potential is less than 1 mV. This ensures that rotation of the bridge balance control RV8 does not give rise to changes in offset voltage.

(2) Apply a 50–1000 Hz square wave of 0.5–1 V to the probe input. Monitor the input and E SIG OUT differentially on an oscilloscope. Adjust RV5 for null (less than 0.001 of the signal). This sets the gain to unity.

(3) Monitor the input and terminal A of SKT 1 differentially while applying the square wave as above. Adjust RV4 for null, to maximize the preamplifier's input resistance. An error of 1% corresponds to an input resistance of 10 G$\Omega$.

(4) Monitor I SIG OUT while applying the square wave to the input as above. Adjust RV1 for null.

(5) Earth the probe input, monitor E SIG OUT and set the output voltage to zero with the offset controls. Interpose a 100 MΩ resistor between the probe input and earth and shunt it with 0.01 $\mu$F to prevent hum pickup. Adjust RV3 for zero voltage, to minimize the leakage current. An error of 1 mV corresponds to a leakage current of 10 pA.

(6) With the 100 MΩ resistor still connected check that closure of Sw4 gives a d.c. shift of about 100 mV.

The entire calibration procedure should be repeated every few months.

# Appendix V: A Simple Current Pump

## Specifications

Maximum output current: ±130 nA into short circuit load; ±65 nA into 100 MΩ load.

Bias current: adjustable from 0 to ±25 nA.

Rise time: less than 200 $\mu$s into 100 MΩ load.

Output resistance: at least $10^{10}$ Ω.

Input characteristics: input resistance 10 kΩ. Output current is proportional to input signal. An input of 10 mV gives 1 nA output current.

Monitor signals: I SIG OUT gives true measurement of output current at a sensitivity of 10 mV nA$^{-1}$. E SIG OUT monitors the voltage applied to the microelectrode.

Ejection polarity: switch-controlled. Ejection polarity can also be controlled by polarity of the input command signal.

## Principle of Operation

In Fig. 56(a) the potential difference between the two ends of $R_s$ is electronically controlled so as to be proportional to the input command signal. If the potential is other than zero, a current flows through $R_s$. The only available pathway for this current is via the microelectrode, since operational amplifier A1 has a very high input impedance. A1 is a unity gain voltage follower; thus the potential at D is equal to that at terminal A. Differential amplifier A2 has an output which is one tenth of the potential difference between B and D; this signal is passed to the I SIG OUT terminal and is compared with the command signal at point

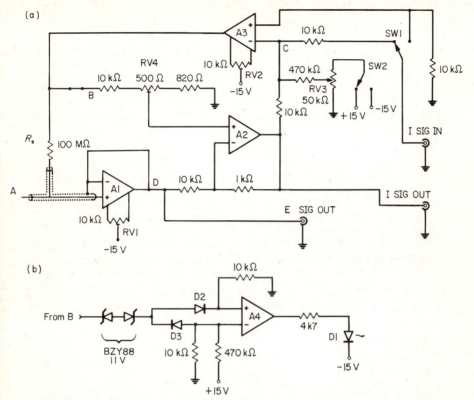

Fig. 56. (a) Circuit diagram of simple current pump. All operational amplifiers are types CA3140. Switches Sw1 (EJECTION POLARITY) and Sw2 (BIAS POLARITY) are single-pole, double-throw with centre OFF. RV3 is a front panel control which varies the steady bias current. (b) Optional overload indicator for current pump. A light-emitting diode D1 glows as a warning whenever the output of amplifier A3 approaches its limits. The zener diodes and amplifier A4 act as a crude window detector; out-of-range voltages are steered to the appropriate input by silicon diodes D2 and D3.

C. Amplifier A3 is the feedback or "current clamp" amplifier which forces the appropriate current through $R_s$. The polarity of the output current is controlled by Sw1 switching the command signal to either the inverting or non-inverting input of A3. Sw1 and also the bias polarity control Sw2 have centre OFF positions.

### Construction

Layout is not critical, except for the wiring at the non-inverting input of A1. To minimize unwanted leakage currents this wiring should be done with stiff wire in mid-air, or on PTFE insulators. It should *not* be run on a printed circuit board. A driven shield must enclose the wiring as shown in Fig. 56(a). Exposed (unscreened) wiring should not exceed a few centimetres in length. The lead to the microelectrode may be a metre or so long, provided it has a driven shield over its entire length. The shield must not be earthed; this would not damage the circuit but would prevent correct operation. The $\pm 15$ V power supply leads should be bypassed to 0 V with capacitors (not shown) to prevent oscillation. Tantalum, 2.2 $\mu$F, 35 V, are suitable.

### Setting Up

Earth the I SIG IN terminal and switch Sw2 OFF. Remove the link at B. Connect a jumper lead from B to A, and apply a 1 kHz, 1 V square wave from a low impedance source between A and earth. Monitor I SIG OUT on an oscilloscope and adjust RV4 for null (less than 1 mV). Replace the link and remove the jumper lead. Earth point A and monitor E SIG OUT. Adjust RV1 for zero voltage (less than 1 mV). Connect a resistor (10–100 M$\Omega$) between A and earth and adjust RV2 for zero voltage.

# Appendix VI: Drug Concentration and Ionophoretic Release

Refer to Chapter 6, Section II for a general discussion of the relation between current flow and ionophoretic drug release. The result derived below has been presented elsewhere without proof (Purves, 1980b). The analysis will be based on three assumptions. (1) The pipette geometry is as shown in Fig. 6 and discussed in Chapter 3, Section IV. (2) The drug electrolyte solution filling the microelectrode is initially at concentration $C_0$, and the external electrolyte solution is at concentration $C_0^*$. (3) All ions under consideration have the same diffusion coefficient $D$ and absolute charge number $|z|$.

The net flux in moles per second of an ion species $i$ is given by the Nernst – Planck equation:

$$j_i = \pi\theta^2 r^2 \frac{z_i F}{RT} \frac{J}{\sigma} DC_i - \pi\theta^2 r^2 D \frac{dC_i}{dr}, \tag{27}$$

where $J = I/\pi\theta^2 r^2$ is the current density. The ionic population at any point along the pipette consists of a mixture of ions deriving originally from both the filling solution and the external medium. In the absence of applied current the total ion concentration is given by equation (4) and the conductivity by the Nernst – Einstein relation (equation (5)):

$$\sigma = \sigma(r) = \frac{2z^2 F^2 D}{RT} C_0^* \left[ \frac{r_0}{r}(1 - w) + w \right], \tag{28}$$

where $w = C_0/C_0^*$. The passage of current through the pipette affects the ratio [drug ions]:[external ions], but can be shown not to alter the total ion concentration. Expression (28) for the conductivity can therefore be used even when the current $I$ is not zero.

135

In the steady state the flux of any ion species along the pipette must be independent of $r$. Substitution of equation (28) into equation (27) and differentiation with respect to $r$ gives

$$0 = -\frac{1}{\pi\theta^2 D}\frac{dj_i}{dr}$$

$$= r^2\frac{d^2 C_i}{dr^2} + \frac{dC_i}{dr}\left(2r - \frac{I}{2zFq_D^*}\frac{rr_0}{r_0(1-w)+rw}\right)$$

$$- C_i\ \frac{I}{2zFq_D^*}\frac{(1-w)r_0^2}{[r_0(1-w)+rw]^2}, \tag{29}$$

where $q_D^* = \pi DC_0^*\theta^2 r_0 \approx \pi DC_0^*\theta a$ is the diffusional leak that the pipette would exhibit if $w = 1$.

Since the ions of interest are drug ions belonging to the filling solution, the appropriate boundary conditions are

$$C_i \to C_0 = wC_0^*, \qquad r \to \infty$$

and

$$C_i = 0, \qquad r = r_0.$$

The solution satisfying equation (29) and the boundary conditions is

$$C_i = \frac{C_0^*\left\{\dfrac{r_0}{r}(1-w) + w - \exp\left(\dfrac{I}{2zFq_D^*(w-1)}\ln\left[\dfrac{r_0}{r}(1-w)+w\right]\right)\right\}}{1 - \exp\left(\left[\dfrac{I}{2zFq_D^*(w-1)}-1\right]\ln w\right)}, \tag{30}$$

which may be verified by differentiation.

The efflux is

$$q = \pi\theta^2 r^2 D\frac{dC_i}{dr}, \qquad r = r_0$$

$$= -\frac{I/2zF + k}{1 - \exp[-(1 + I/2zFk)\ln w]}, \tag{31}$$

where

$$k = q_D^*(1-w).$$

This result is plotted in Fig. 43.

# References

Adrian, R. H. (1956). The effect of internal and external potassium concentration on the membrane potential of frog muscle. *J. Physiol., Lond.* **133**, 631–658.

Agin, D. P. (1969). Electrochemical properties of glass microelectrodes. In "Glass Microelectrodes" (M. Lavallée, O. F. Schanne and N. C. Hébert, eds), pp. 62–75. Wiley, New York.

Agin, D. and Holtzman, D. (1966). Glass microelectrodes: the origin and elimination of tip potentials. *Nature, Lond.* **211**, 1194–1195.

Alexander, J. and Nastuk, W. L. (1953). An instrument for the production of microelectrodes used in electrophysiological studies. *Rev. scient. Instrum.* **24**, 528–531.

Amatniek, E. (1958). Measurement of bioelectric potentials with microelectrodes and neutralized input capacity amplifiers. *I.R.E. Trans. med. Electron.* **PGME-10**, 3–14.

Araki, T. and Otani, T. (1955). Response of single motoneurons to direct stimulation in toad's spinal cord. *J. Neurophysiol.* **18**, 472–485.

Bach-y-Rita, P. and Ito, F. (1966). *In vivo* studies on fast and slow muscle fibers in cat extraocular muscles. *J. gen. Physiol.* **49**, 1177–1198.

Bader, C. R., MacLeish, P. R. and Schwartz, E. A. (1979). A voltage-clamp study of the light response in solitary rods of the tiger salamander. *J. Physiol., Lond.* **296**, 1–26.

Baldwin, D. J. (1980a). Dry bevelling of micropipette electrodes. *J. Neuroscience Methods* **2**, 153–161.

Baldwin, D. J. (1980b). Non-destructive electron microscopic examination with rotation of beveled micropipette electrode tips. *J. Neuroscience Methods* **2**, 163–167.

Barrett, J. N. and Graubard, K. (1970). Fluorescent staining of cat motoneurons *in vivo* with beveled micropipettes. *Brain Res.* **18**, 565–568.

Bils, R. F. and Lavallée, M. (1964). Measurement of glass microelectrodes. *Experientia* **20**, 231–232.

Bloom, F. E. (1974). To spritz or not spritz: the doubtful value of aimless iontophoresis. *Life Sci.* **14**, 1819–1834.

Bockris, J. O'M. and Reddy, A. K. N. (1970). "Modern Electrochemistry", Vol. 1. Plenum Press, New York.

Boev, K. and Golenhofen, K. (1974). Sucrose-gap technique with pressed rubber membranes. *Pflüger's Arch.* **349**, 277–283.

Brown, K. T. and Flaming, D. G. (1974). Bevelling of fine micropipette tips by a rapid precision method. *Science, N.Y.* **185**, 693–695.

Brown, K. T. and Flaming, D. G. (1975). Instrumentation and technique for bevelling fine micropipette electrodes. *Brain Res.* **86**, 172–180.

Brown, K. T. and Flaming, D. G. (1977). New microelectrode techniques for intracellular work in small cells. *Neuroscience* **2**, 813–827.

Brown, K. T. and Flaming, D. G. (1979). Technique for precision beveling of relatively large micropipettes. *J. Neurosci. Methods* **1**, 25–35.

Brown, P. B., Maxfield, B. W. and Moraff, H. (1973). "Electronics for Neurobiologists". M.I.T. Press, Cambridge, Mass.

Bureš, J., Petráň, M. and Zachar, J. (1967). "Electrophysiological Methods in Biological Research". Academic Press, New York and London.

Byzov, A. L. and Chernyshov, V. I. (1961). Automatic device for manufacturing microelectrodes. *Biophysics* **6**, 536–541.

Caldwell, P. C. and Downing, A. C. (1955). Preparation of capillary microelectrodes. *J. Physiol., Lond.* **128**, 31P.

Cerf, J. A. and Cerf, E. (1974). A holder for rapid filling of micropipette electrodes by centrifugal action. *Pflüger's Arch.* **349**, 87–90.

Chowdhury, T. K. (1969*a*). Fabrication of extremely fine glass micropipette electrodes. *J. scient. Instrum.* **2**, 1087–1090.

Chowdhury, T. K. (1969*b*). Techniques of intracellular micro-injection. In "Glass Microelectrodes" (M. Lavallée, O. F. Schanne and N. C. Hébert, eds), pp. 404–423, Wiley, New York.

Coburn, R. F., Ohba, M. and Tomita, T. (1975). Recording of intracellular electrical activity with the sucrose gap method. In "Methods in Pharmacology", Vol. 3, "Smooth Muscle" (E. E. Daniel and D. M. Paton, eds), pp. 231–245. Plenum Press, New York.

Colburn, T. R. and Schwartz, E. A. (1972). Linear voltage control of current passed through a micropipette with variable resistance. *Med. biol. Engng.* **10**, 504–509.

Coombs, J. S., Eccles, J. C. and Fatt, P. (1955). The specific ionic conductances and the ionic movements that produce the inhibitory postsynaptic potential *J. Physiol., Lond.* **130**, 326–373.

Corson, D. W., Goodman, S. and Fein, A. (1979). An adaptation of the jet stream microelectrode beveler. *Science, N.Y.* **205**, 1302.

Crain, S. M. (1956). Resting and action potentials of cultured chick embryo spinal ganglion cells. *J. comp. Neurol.* **104**, 285–329.

Crank, J. (1975). "The Mathematics of Diffusion", 2nd edn. University Press, Oxford.

Curtis, D. R. (1964). Microelectrophoresis. In "Physical Techniques in Biological Research", Vol. V, "Electrophysiological Methods", Part A (W. L. Nastuk, ed.), pp. 144–190, Academic Press, New York and London.

DeFelice, L. J. and Firth, D. R. (1971). Spontaneous voltage fluctuations in glass microelectrodes. *I.E.E.E. Trans. biomed. Electron.* **BME-18**, 339–351.

Delgado, J. M. R. (1964). Electrodes for extracellular recording and stimulation. In "Physical Techniques in Biological Research", Vol. V, "Electrophysiological Methods", Part A (W. L. Nastuk, ed.), pp. 88–143. Academic Press, New York and London.

Dichter, M. A. (1973). Intracellular single unit recording. In "Methods in Physiological Psychology", Vol. 1A (R. F. Thompson and M. M. Patterson, eds), pp. 3–21. Academic Press, New York and London.

Donaldson, P. E. K. (1958). "Electronic Apparatus for Biological Research". Butterworths, London.

Dreyer, F. and Peper, K. (1974). Iontophoretic application of acetylcholine: advantages of high resistance micropipettes in connection with an electronic current pump. *Pflüger's Arch.* **348**, 263–272.

El-Badry, H. M. (1963). "Micromanipulators and Micromanipulation". Springer-Verlag, Vienna.

Engel, E., Barcilon, V. and Eisenberg, R. S. (1972). The interpretation of current – voltage relations recorded from a spherical cell with a single microelectrode. *Biophys. J.* **12**, 384–403.

Ensor, D. R. (1979). A new moving-coil microelectrode puller. *J. Neurosci. Methods* **1**, 95–105.

Fatt, P. (1961). Intracellular microelectrodes. In "Methods in Medical Research", Vol. 9 (J. H. Quastel, ed.), pp. 381–404. Year Book Medical Publishers, Chicago.

Fatt, P. and Katz, B. (1951). An analysis of the end-plate potential recorded with an intracellular electrode. *J. Physiol., Lond.* **115**, 320–370.

Fein, H. (1966). Passing current through recording glass micropipette electrodes. *I.E.E.E. Trans. biomed. Electron.* **BME-13**, 211–212.

Firth, D. R. and DeFelice, L. J. (1971). Electrical resistance and volume flow in glass microelectrodes. *Can. J. Physiol. Pharmac.* **49**, 436–447.

Forman, D. S. and Cruce, W. L. R. (1972). Darkfield electron microscopy: a simple method for the examination of glass microelectrodes in the electron microscope. *Electroenceph. clin. Neurophysiol.* **33**, 427–429.

Frank, K. and Becker, M. C. (1964). Microelectrodes for recording and stimulation. In "Physical Techniques in Biological Research", Vol. V, "Electrophysiological Methods" Part A (W. L. Nastuk, ed.), pp. 22–87. Academic Press, New York and London.

Frank, K. and Fuortes, M. G. F. (1956). Stimulation of spinal motoneurons with intracellular electrodes. *J. Physiol., Lond.* **134**, 451–470.

Fry, D. M. (1975). A scanning electron microscope method for the examination of glass microelectrode tips either before or after use. *Experientia* **31**, 695–697.

Gage, P. W. and Eisenberg, R. S. (1969). Capacitance of the surface and transverse tubular membrane of frog sartorius muscle fibers. *J. gen. Physiol.* **53**, 265–278.

Geddes, L. A. (1972). "Electrodes and the Measurement of Bioelectric Events". Wiley, New York.

Globus, A. (1973). Iontophoretic injection techniques. In "Methods in Physiological Psychology", Vol. 1A. (R. F. Thompson and M. M. Patterson, eds), pp. 23–38, Academic Press, New York and London.

Gordon, J. E. (1976). "The New Science of Strong Materials", 2nd edn. Penguin, Harmondsworth.

Gotow, T., Ohba, M. and Tomita, T. (1977). Tip potential and resistance of microelectrodes filled with KCl solution by boiling and nonboiling methods. *I.E.E.E. Trans. biomed. Electron.* **BME-24**, 366–371.

Guld, C. (1962). Cathode follower and negative capacitance as high input impedance circuits. *Proc. I.R.E.* **50**, 1912–1927.

Guyenet, P. G., Mroz, E. A., Aghajanian, G. K. and Leeman, S. E. (1979). Delayed iontophoretic ejection of substance P from glass micropipettes: correlation with time-course of neuronal excitation *in vivo*. *Neuropharmacology* **18**, 553–558.

Henderson, R. M. (1967). Preparation of combined sodium-sensitive and reference microelectrodes. *J. appl. Physiol.* **22**, 1179–1181.

Hill-Smith, I. and Purves, R. D. (1978). Synaptic delay in the heart: an ionophoretic study. *J. Physiol., Lond.* **279**, 31–54.

Hironaka, T. and Morimoto, S. (1979). The resting potential of frog sartorius muscle. *J. Physiol., Lond.* **297**, 1–8.

Hogg, B. M., Goss, C. M. and Cole, K. S. (1934). Potentials in embryo rat heart cultures. *Proc. Soc. exp. Biol.* **32**, 304–307.

Hudspeth, A. J. and Corey, D. P. (1978). Controlled bending of high-resistance glass microelectrodes. *Am. J. Physiol.* **234**, C56–C57.

Isenberg, G. (1979). Risk and advantages of using strongly beveled microelectrodes for electrophysiological studies in cardiac Purkinje fibers. *Pflüger's Arch.* **380**, 91–98.

Ito, M., Kostyuk, P. G. and Oshima, T. (1962). Further study on anion permeability of inhibitory post-synaptic membrane of cat motoneurones. *J. Physiol., Lond.* **164**, 150–156.

Ives, D. J. G. and Janz, G. J. (1961). "Reference Electrodes". Academic Press, New York and London.

Jacobson, S. L. and Mealing, G. A. R. (1980). A method for producing very low resistance micropipettes for intracellular recording. *Electroenceph. clin. Neurophysiol.* **48**, 106–108.

Janz, G. J. and Ives, D. J. G. (1968). Silver, silver chloride electrodes. *Ann. N.Y. Acad. Sci.* **148**, 210–221.

Källström, Y. and Lindström, S. (1978). A simple device for pressure injections of horseradish peroxidase into small central neurones. *Brain Res.* **156**, 102–105.

Kao, C. Y. (1954). A method of making prefilled microelectrodes. *Science, N.Y.* **119**, 846–847.

Kater, S. B. and Nicholson, C. (1973). "Intracellular Staining in Neurobiology". Springer, Berlin.

Kelly, J. S. (1975). Microiontophoretic application of drugs onto single neurons. In "Handbook of Psychopharmacology", Vol. 2 (L. L. Iversen, S. H. Iversen and S. D. Snyder, eds), pp. 29–67. Plenum Press, New York.

Kelly, J. S., Simmonds, M. A. and Straughan, D. W. (1975). Microelectrode techniques. In "Methods in Brain Research" (P. B. Bradley, ed.), Chap. 7, pp. 333–377. Wiley, London.

Kennard, D. W. (1958). In "Electronic Apparatus for Biological Research" (P. K. Donaldson, ed.), Chap. 35, pp. 534–567. Academic Press, New York and London.

Kite, L. V. (1974). "An Introduction to Linear Electric Circuits". Longmans, Harlow, Essex.

Kopac, M. J. (1964). Micromanipulators: principles of design, operation and application. In "Physical Techniques in Biological Research", Vol. V, "Electrophysiological Methods", Part A (W. L. Nastuk, ed.), pp. 191–233. Academic Press, New York and London.

Kostyuk, P. G. and Krishtal, O. A. (1977). Separation of sodium and calcium currents in the somatic membrane of mollusc neurones. *J. Physiol., Lond.* **270**, 545–568.

Kripke, B. R. and Ogden, T. E. (1974). A technique for beveling fine micropipettes. *Electroenceph. clin. Neurophysiol.* **136**, 323–326.

Krischer, C. C. (1969a). Theoretical treatment of ohmic and rectifying properties of electrolyte filled micropipettes. *Z. Naturforsch. B.* **24**, 151–155.

Krischer, C. C. (1969*b*). Current voltage measurements of electrolyte filled microelectrodes with ohmic and rectifying properties. *Z. Naturforsch. B.* **24**, 156–161.

Krnjević, K. (1971). Microiontophoresis. In "Methods of Neurochemistry", Vol. 1 (R. Fried, ed.), pp. 129–172. Dekker, New York.

Krnjević, K., Mitchell, J. F. and Szerb, J. C. (1963). Determination of iontophoretic release of acetylcholine from micropipettes. *J. Physiol., Lond.* **165**, 421–436.

Küchler, G. (1964). Zur Frage der Ubertragungseigenschaften von Glasmikroelektroden bei der intracellulären Membranpotentialmessung. *Pflüger's Arch.* **280**, 210–223.

Kurella, G. A. (1958). A method of producing intracellular microelectrodes. *Biophysics* **3**, 228–231.

Kuriyama, H. and Ito, Y. (1975). Recording of intracellular electrical activity with microelectrodes. In "Methods in Pharmacology", Vol. III, "Smooth Muscle" (E. E. Daniel and D. M. Paton, eds), pp. 201–230. Plenum Press, New York.

Lanthier, R. and Schanne, O. (1966). Change of microelectrode resistance in solutions of different resistivities. *Naturwissenschaften* **53**, 430–431.

Lassen, U. V. and Sten-Knudsen, O. (1968). Direct measurements of membrane potentials and membrane resistance of human red cells. *J. Physiol., Lond.* **195**, 681–696.

Lavallée, M. and Szabo, G. (1969). The effect of glass surface conductivity phenomena on the tip potential of micropipette electrodes. In "Glass Microelectrodes" (M. Lavallée, O. F. Schanne and N. C. Hébert, eds), pp. 95–110. Wiley, New York.

Lavallée, M., Schanne, O. F. and Hébert, N. C. (1969). "Glass Microelectrodes". Wiley, New York.

Lederer, W. J., Spindler, A. J. and Eisner, D. A. (1979). Thick slurry bevelling. A new technique for bevelling extremely fine microelectrodes and micropipettes. *Pflüger's Arch.* **381**, 287–288.

Lee, K. S., Akaike, N. and Brown, A. M. (1980). The suction pipette method for internal perfusion and voltage clamp of small excitable cells. *J. Neurosci. Method* **2**, 51–78.

Lester, H. A., Nass, M. M., Krouse, M. E., Nerbonne, J. M., Erlanger, B. F. and Wassermann, N. H. (1980). Electrophysiological experiments with photoisomerisable cholinergic compounds. *Ann. N.Y. Acad. Sci.* **346**, 475–490.

Levine, L. (1966). Tip potentials in microelectrodes filled by boiling under reduced pressure. *Experientia* **22**, 559–560.

Ling, G. and Gerard, R. W. (1949). The normal membrane potential of frog sartorius fibers. *J. cell comp. Physiol.* **34**, 383–396.

Livingston, L. G. and Duggar, B. M. (1934). Experimental procedures in a study of the location and concentration within the host cell of tobacco mosaic. *Biol. Bull. Mar. Biol. Lab., Woods Hole* **67**, 504–512.

McCaman, R. E., McKenna, D. G. and Ono, J. K. (1977). A pressure system for intracellular and extracellular ejections of picoliter volumes. *Brain Res.* **136**, 141–147.

Martin, A. R. and Pilar, G. (1963). Dual mode of transmission in the avian ciliary ganglion. *J. Physiol., Lond.* **168**, 443–463.

Merickel, M. (1980). Design of a single electrode voltage clamp. *J. Neurosci. Methods* **2**, 87–96.

Moore, E. N. and Bloom, M. (1969). A method for intracellular stimulation and recording using a single microelectrode. *J. appl. Physiol.* **27**, 734–735.

Moore, J. W. and Gebhart, J. H. (1962). Stabilized wide-band potentiometric preamplifiers. *Proc. Inst. Radio Engrs.* **50**, 1928–1941.

Morad, M. and Salama, G. (1979). Optical probes of membrane potential in heart muscle. *J. Physiol., Lond.* **292**, 267–295.

Motchenbacher, C. D. and Fitchen, F. C. (1973). "Low-Noise Electronic Design". Wiley, New York.

Muijser, H. (1979). A microelectrode amplifier with an infinite resistance current source for intracellular measurements of membrane potential and resistance changes under current clamp. *Experientia* **35**, 912–913.

Mullins, L. J. and Noda, K. (1963). The influence of sodium-free solutions on the membrane potential of frog muscle fibers. *J. gen. Physiol.* **47**, 117–132.

Nastuk, W. L. (1953a). The electrical activity of the muscle cell membrane at the neuromuscular junction. *J. cell. comp. Physiol.* **42**, 249–272.

Nastuk, W. L. (1953b). Membrane potential changes at a single muscle end-plate produced by transitory application of acetylcholine with an electrically controlled microjet. *Fedn Proc. Fedn Am. Socs exp. Biol.* **12**, 102.

Nastuk, W. L. (1964). "Physical Techniques in Biological Research", Vols V and VI. Academic Press, New York and London.

Nastuk, W. L. and Hodgkin, A. L. (1950). The electrical activity of single muscle fibres. *J. cell. comp. Physiol.* **35**, 39–73.

Neale, E. A., MacDonald, R. L. and Nelson, P. G. (1978). Intracellular horseradish peroxidase injection for correlation of light and electron microscopic anatomy with synaptic physiology. *Brain Res.* **152**, 265–282.

Ogden, T. E., Citron, M. C. and Pierantoni, R. (1978). The jet stream microbeveler: an inexpensive way to bevel glass micropipettes. *Science, N.Y.* **201**, 469–470.

Okada, Y. and Inouye, A. (1975). Tip potential and fixed charges on the glass wall of microelectrode. *Experientia* **31**, 545–546.

Ott, H. W. (1976). "Noise Reduction Techniques in Electronic Systems". Wiley, New York.

Peskoff, A. and Eisenberg, R. S. (1975). The time-dependent potential in a spherical cell using matched asymptotic expansions. *J. math. Biol.* **2**, 277–300.

Plamondon, R., Gagné, S. and Poussart, D. (1979). Low resistance and tip potential of microelectrode: improvement through a new filling method. *Vision Res.* **16**, 1355–1357.

Proenza, L. M. and Morton, R. E. (1975). A device for beveling fine micropipettes. *Physiol. Behav.* **14**, 511–513.

Purves, R. D. (1976). Microelectrodes in spherical cells. *J. theor. Biol.* **63**, 225–228.

Purves, R. D. (1977). The release of drugs from iontophoretic pipettes. *J. theor. Biol.* **66**, 789–798.

Purves, R. D. (1979). The physics of iontophoretic pipettes. *J. Neurosci. Methods* **1**, 165–178.

Purves, R. D. (1980a). The mechanics of pulling a glass micropipette. *Biophys. J.* **29**, 523–530.

Purves, R. D. (1980b). Effect of drug concentration on release from ionophoretic pipettes. *J. Physiol., Lond.* **300**, 72P–73P.

Rae, J. L. and Germer, H. A. (1974). Junction potentials in the crystalline lens. *J. appl. Physiol.* **37**, 464–467.

Riemer, J., Mayer, C.-J. and Ulbrecht, G. (1974). Determination of membrane potential in smooth muscle cells using microelectrodes with reduced tip potential. *Pflüger's Arch.* **349**, 267–275.

Riggs, D. S. (1963). "The Mathematical Approach to Physiological Problems". Williams & Wilkins, Baltimore.

Robinson, G. R. and Scott, B. I. H. (1973). A new method of estimating micropipette tip diameter. *Experientia* **29**, 1039–1040.

Rosenberg, M. (1973). A comparison of chloride and citrate filled microelectrodes for d-c recording. *J. appl. Physiol.* **35**, 166–168.

Rubio, R. and Zubieta, G. (1961). The variation of the electric resistance of microelectrodes during the flow of current. *Acta physiol. latino-am.* **11**, 91–94.

Rush, S., Lepeschkin, E. and Brooks, H. O. (1968). Electrical and thermal properties of double-barrelled ultra microelectrodes. *I.E.E.E. Trans. biomed. Electron.* **BME-15**, 80–93.

Sakai, M., Swartz, B. E. and Woody, C. D. (1979). Controlled release of pharmacological agents: measurements of volume ejected *in vitro* through fine tipped glass microelectrodes by pressure. *Neuropharmacology* **18**, 209–213.

Sato, K. (1977). Modifications of glass microelectrodes: a self-filling and semifloating glass microelectrode. *Am. J. Physiol.* **232**, C207–C210.

Schanne, O. F. (1969). Measurement of cytoplasmic resistivity by means of the glass microelectrode. In "Glass Microelectrodes" (M. Lavallée, O. F Schanne and N. C. Hébert, eds), pp. 299–321. Wiley, New York.

Schanne, O., Kawata, H., Schäfer, B. and Lavallée, M. (1966). A study on the electrical resistance of frog sartorius muscle. *J. gen. Physiol.* **49**, 897–912.

Schanne, O. F., Lavallée, M., Laprade, R. and Gagné, S. (1968). Electrical properties of glass microelectrodes. *Proc. I.E.E.E.* **56**, 1072–1082.

Schoenfeld, R. L. (1962). Bandwidth limits for neutralized input capacity amplifiers. *Proc. Inst. Radio Engrs.* **50**, 1942–1950.

Shaw, D. J. (1970). "Introduction to Colloid and Surface Chemistry", 2nd edn. Butterworths, London.

Shaw, M. L. and Lee, D. R. (1973). Micropipette sharpener with audio and hydraulic readouts. *J. appl. Physiol.* **34**, 523–524.

Smith, J. I. (1971). "Modern Operational Circuit Design". Wiley, New York.

Snell, F. M. (1969). Some electrical properties of fine-tipped pipette microelectrodes. In "Glass Microelectrodes" (M. Lavallée, O. F. Schanne and N. C. Hébert, eds), pp. 111–123. Wiley, New York.

Spindler, A. J. (1979). Miniaturized intracellular pressure injector. *J. Physiol., Lond.* **296**, 4P–5P.

Stark, P. A. (1970). "Introduction to Numerical Methods". Macmillan, London.

Stewart, W. W. (1978). Functional connections between cells as revealed by dye-coupling with a highly fluorescent napthalimide tracer. *Cell* **14**, 741–759.

Strong, P. (1970). "Biophysical Measurements". Tektronix Inc. Beaverton, Oregon.

Tasaki, I. and Singer, I. (1968). Some problems involved in electric measurements of biological systems. *Ann. N.Y. Acad. Sci.* **148**, 36–53.

Tasaki, I., Polley, E. H. and Orrego, F. (1954). Action potentials from individual elements in cat geniculate and striate cortex. *J. Neurophysiol.* **17**, 454–474.

Tasaki, K., Tsukahara, Y., Ito, S., Wayner, M. J. and Yu, W. Y. (1968). A simple, direct and rapid method for filling microelectrodes. *Physiol. Behav.* **3**, 1009–1010.

Tauchi, M. and Kikuchi, R. (1977). A simple method for bevelling micropipettes for intracellular recording and current injection. *Pflüger's Arch.* **368**, 153–155.

Taylor, R. E. (1953). Quoted in Jenerick, H. and Gerard, R. W. (1953). Membrane potential and threshold of single muscle fibres. *J. cell. comp. Physiol.* **42**, 79–102.

Thomas, M. V. (1977). Microelectrode amplifier with improved method of input capacitance neutralization. *Med. biol. Engng Comput.* **15**, 450–454.

Thomas, R. C. (1972). Intracellular sodium activity and the sodium pump in snail neurones. *J. Physiol., Lond.* **220**, 55–71.

Thomas, R. C. (1975). A floating current clamp for intracellular injection of salts by interbarrel iontophoresis. *J. Physiol., Lond.* **245**, 20P–22P.

Thomas, R. C. (1977). The role of bicarbonate, chloride and sodium ions in the regulation of intracellular pH in snail neurones. *J. Physiol., Lond.* **273**, 317–338.

Thomas, R. C. (1978). "Ion-Sensitive Microelectrodes: How to Make and Use Them". Academic Press, London and New York.

Ujec, E., Vít, Z., Vyskočil, F. and Králík, O. (1973). Analysis of geometrical and electrical parameters of tips of glass microelectrodes. *Physiol. bohemoslov.* **22**, 329–336.

Unwin, D. M. and Moreton, R. B. (1974). A microelectrode amplifier with f.e.t. input and facility for input capacity compensation. *Med. biol. Engng* **12**, 344–347.

Vyskočil, F. and Melichar, I. (1972). A simple method of filling microelectrodes by vapour condensation. *Physiol. bohemoslov.* **21**, 451–452.

Wann, K. T. and Goldsmith, M. W. (1972). Reduction of tip potential artefacts in microelectrode measurements. *Nature phys. Sci.* **238**, 44–45.

Willis, J. A., Myers, P. R. and Carpenter, D. O. (1977). An ionophoretic module which controls electroosmosis. *J. electrophysiol. Tech.* **6**, 34–41.

Wilson, W. A. and Goldner, M. M. (1975). Voltage clamping with a single microelectrode. *J. Neurobiol.* **6**, 411–422.

Winsbury, G. I. (1956). Machine for the production of microelectrodes. *Rev. scient. Instrum.* **27**, 514–516.

Wolbarsht, M. L. (1964). Interference and its elimination. In "Physical Techniques in Biological Research", Vol. V, "Electrophysiological Methods", Part A (W. L. Nastuk, ed.), pp. 353–372. Academic Press, New York and London.

Woodbury, J. W. and Brady, A. J. (1956). Intracellular recording from moving tissues with a flexibly mounted ultramicroelectrode. *Science, N.Y.* **123**, 100–101.

Young, S. (1973). "Electronics in the Life Sciences". Macmillan, London.

Zeuthen, L. (1971). A method to fill glass microelectrodes by local heating. *Acta. physiol. scand.* **81**, 141–142.

Zimmerer, R. P. (1973). A simple, rapid, aseptic method for filling micropipettes by centrifugation. *Expl Cell Res.* **78**, 250–251.

# Index

Ågar bridge, 53
Amplifier, *see also* Preamplifier
  operational, 43, 113–115
  symbols, 113
Artefact from stimulus, 66

Baseplate, 103, 106
Batteries, 43, 65, 75, 120, 121
Bevelling, 24–25, 90
Breakaway box, 81–82
Bridge for current injection, 82–91

Cable, capacitance, 32, 122
  faults, 75
Calibration, 73–74
Capacitance, definition, 110
  of input circuit, 31–33, 45, 48, 86
  negative, 47–49, 86, 88, 91
Cathodal screen, *see* Driven shield
Cathode follower, 5, 44
Chloriding silver, 51
Current, balancing, 99, 101
  monitors, 79–81
  pumps, 78, 99, 114–115, 132–134
  sources, 76–79
Current-to-voltage converter, 80–81, 115

Driven shield, 45–47, 70, 79, 122
Dyes sensitive to voltage, 12

Earth loop, 59–61, 65
Earthing, 55–61
Electron microscopy, 27
Electronics, 107–115
Electro-osmosis, 37, 94, 118
Extracellular recording, 10, 110

Faraday's law, 93
Field effect transistor, 43
Filling methods, 17–24
Filter, low-pass, 45–46, 68, 72–73
  notch, 73

Glass, tubing, 13–14
  fibre, 23
Grounding, *see* Earthing

Hittorf's law, 93
Hum, 58, 61–65

Impalement, 8–9
Impedance, 111
Indifferent electrode, 4, 51–54
Input resistance of amplifier, 5, 40
Interference, 57–67
Intracellular ionophoresis, 100–101
Ionophoresis, 3, 92–102
  physics, 92–98, 135–136

Leakage current, 5, 40
Liquid junction potential, 9, 16, 33, 53

Microelectrode, 3
  capacitance, 31–33
  resistance, *see* Resistance
Micromanipulator, 5–6, 105
Micropipette, 3
  puller, 14–15
  tip diameter, 26–27
Microscope, 8

Negative capacitance, *see* Capacitance
Nernst–Planck equation, 116

Noise, 67–71
Non-linear electrode properties,
    35–38, 116–119

Offset voltage, 42, 52, 55, 121, 128
Operational amplifier, *see* Amplifier
Oscilloscope, 6

Preamplifier, 5, 39–43
    circuits, 123, 126–127
Pressure ejection, 94, 102

Reciprocity theorem, 109, 110
Resistance of microelectrode, 28–29
    measurement, 71–72
    theory, 29–31, 35–38, 116–119
Resting potential, 9–10
Retaining current, 94, 95, 96, 102
Rise time, 41, 73

Safety, 55
Salt bridge, 3, 10

Screening, 46, 57–58
Shunt, 108
Silver/silver chloride electrode, 50–51
Stimulus artefact, 66
Sucrose gap, 10
Superposition theorem, 110

Tape recorder, 7
Time constant, 112
    of input circuit, 45, 88, 91
Tip potential, 9, 33–35
Transport number, 93
Trouble-shooting, 74–75

Vibration, 103–106
Virtual earth, 115
Voltage clamp with single electrode,
    91
Voltage follower, 44, 79, 114, 120

Wheatstone bridge, *see* Bridge

Zener diode, 122